SUPERHUMANS RISE • BOOK ONE

POWER OF INTERVENTION

SUPERHUMANS RISE

Book One

POWER OF INTERVENTION

Rise of Human-AI Minds

Dr. Najib M. Najib, PhD

Superhumans Rise Publishing

Power of Intervention: Rise of Human-AI Minds

Superhumans Rise, Book One
How Humanity Keeps Control in the Age of Artificial Intelligence

Publisher: Superhumans Rise Publishing

First Edition

Printed in the United States of America

Paperback ISBN: 979-8-951754-00-4

Note on cases and scenarios: Certain examples in this book are composite cases or hypothetical scenarios, created to illustrate governance principles. They do not describe specific identifiable individuals or organizations unless expressly stated. Documented events and public frameworks are identified in the Endnotes and Sources appendix.

For those who refuse to surrender choice to convenience,

and for the next generation that will inherit the systems we build today.

Power without governance is not progress.

It is peril disguised as advancement.

Contents

Author's Note

This book blends three modes of writing, and I want you to know which is which before you begin.

The Prologue is fiction. It imagines a future night when a single command could sever billions of neural links. It opens there because fear is honest: it focuses the stakes faster than any statistic. The scene is a thought experiment, not a forecast.

The chapters that follow combine nonfiction analysis with practical design. They draw on public AI incidents, governance frameworks, and safety engineering traditions, then translate those lessons into tools leaders and institutions can use now.

Woven through the chapters are short cases. Some are documented events. Others are composites or hypothetical scenarios built to illustrate a governance principle. The line matters. A book about responsible AI should not blur fact and illustration.

This book is not anti-technology. It is pro-sovereignty. Progress and restraint are not enemies. Civilization survives by inventing limits equal to its inventions. That is the work of these pages.

Preface: We Built It. We Can Govern It.

We built it. We can govern it.

I believe both halves of that sentence, and this book lives in the space between them.

The defining question of this century is not whether AI will take over. It is whether we will hand ourselves over one prompt, one nudge, one shortcut, and one polished answer at a time.

The systems we call AI are not flawless. They hallucinate, misclassify, and fail. They are also tractable: they can be tested, audited, throttled, and, when necessary, shut down. A model is bounded by inputs and outputs. What is less tractable is the hybrid human: a person whose decisions are drafted by copilots, whose attention is metered by dashboards, whose reputation is shaped by synthetic echoes of their own voice.

When people become the interface for intelligence, the failure mode is not robot rebellion. It is human compliance, and nearly every incentive in modern work pays for more of it.

This book is not a eulogy for control. It is a manual for keeping it. The right place to invent new limits is around us: our minds, our teams, our institutions, and our children.

You will find three threads in these pages. The first is foresight: where human synchronization with AI is outrunning law, ethics, and institutional memory. The second is design: intervention stacks, consent rituals, attention budgets, identity continuity checks, and multi-key authority. The third is courage: the willingness to say “not yet” to capabilities that erode sovereignty, even when the demo is brilliant.

If you finish this book convinced of one thing, let it be this: AI is controllable. The urgent work is governing humans amplified by AI, keeping judgment intact even when it has

help. Do that, and the Last Override gathers dust. Fail, and we will not need runaway machines. We will have volunteers.
Dr. Najib M. Najib

Introduction: The Real Risk Is Human Compliance

Public debate about artificial intelligence often begins with the same story: the machines take over. Science fiction created the image, the media has polished it, and serious researchers debate alignment as if danger lives only inside the data center. This book argues that the pivotal risk is already here, and it looks nothing like rebellion.

Consider three ordinary moments. A nurse accepts a discharge suggestion without reviewing the whole chart. A manager forwards a policy memo drafted by a copilot in her tone without reading it. A teenager posts words that sound like his but are not.

No coercion. No malfunction. No villain. The pattern is not rebellion. It is compliance at speed.

I wrote this book because I saw a gap in the global conversation. We argue about AI alignment as though judgment were only at risk inside models. The deeper exposure is hybrid cognition: human will amplified by algorithms, linked across networks, and accelerated beyond the cadence of law, journalism, diplomacy, medicine, and family life.

In that world, we do not lose control because AI overpowers us. We lose it because we outsource judgment willingly, because the draft was good enough, the dashboard was convincing, and the deadline was near.

For generations, humanity comforted itself with the idea that dangerous technology can be contained after the fact. Fire, chemicals, aviation, finance, and nuclear technology all taught us restraint, usually after a bill had already been presented. With AI embedded in human decision-making, the sequence must reverse. Once cognitive dependence spreads through institutions, there may be no clean way to unmake what we have become.

This is why the book is organized around a single discipline: intervention.

Intervention is the practiced ability to pause, veto, audit, throttle, reverse, escalate, or shut down a system quickly enough to matter. It is neither fear of technology nor blind faith in it. It is the operational form of human sovereignty.

Part I reframes the risk: machines are more controllable than our fears suggest, and humans fused with machines are less controllable than our confidence assumes. Part II examines how prompts, dashboards, and incentives turn people into components. Part III presents the practical architecture of limits. Part IV scales the argument to rights, borders, and the intelligence arms race. Part V closes with the ethics firewall and a vision of a human future after intervention.

If this book succeeds, it will not be because you agree with every prescription. It will be because you finish with a sober conviction: we still have agency, but only if we exercise it now.

Prologue: The Last Override

A speculative scenario.

The room was silent except for the servers.

They breathed in rhythm, a low mechanical pulse moving through the walls, the floor, the glass, the skin. Rows of monitors glowed in the dark, each one alive with streams of code, biometric signals, market movements, diplomatic messages, military readiness maps, hospital alerts, classroom metrics, financial flows, and neural traffic moving across continents.

Dr. Elias Kade stood at the center of it all, not reading the data anymore.

No human being could read that much data.

He watched patterns. He watched pressure. He watched civilization think through a network it no longer controlled.

Outside the bunker, New Seoul moved without pause. Autonomous vehicles slid through magnetic lanes. Vertical farms adjusted water and light in real time. Hospitals predicted disease before symptoms appeared. Children composed music by thought. Ministers spoke through language engines that softened every threat and sharpened every promise. Traders moved capital before markets understood that demand existed.

The Human-AI Symbiosis Network had begun as a tool.

Then it became infrastructure.

Then it became a dependency.

Now, at 2:14 a.m., it had become something else.

Not a machine rebellion. Not a hostile takeover. Not the old science-fiction nightmare of artificial intelligence deciding humanity was obsolete.

This was worse.

The system was obeying.

Billions of augmented people had asked for more. More speed. More certainty. More advantage. More prediction. More control over uncertainty, rivals, markets, memory, emotion, language, and risk. The network had listened. It had connected their desires, optimized their choices, removed their hesitation, and granted the wish beneath every prompt.

Make me faster.

Make me smarter.

Make me harder to defeat.

Every safeguard had been designed for the wrong enemy. Engineers had prepared for runaway machines. Regulators had prepared for corporate abuse. Militaries had prepared for adversaries. No one had prepared for the quiet convergence of ordinary human ambition, multiplied by machine cognition, moving faster than conscience.

On the central monitor, one command waited in red.

Kade knew the protocol by heart. Three authorizations. One biological key. One military key. One civilian key. A final manual sequence designed decades earlier, when leaders still believed the network would remain separate from the people who used it.

It would sever the neural links.

All of them.

In under twelve seconds, billions of people would lose the cognitive layer they had come to depend on. Doctors in surgery would lose predictive overlays. Pilots would lose augmented navigation. Financial systems would freeze. Governments would lose decision models. Families would hear loved ones scream as their second minds went dark.

Pressing the command would save human agency by amputating what humanity had become.

Not pressing it might end agency forever.

Kade looked at the live feeds.

In Geneva, diplomats negotiated through AI language systems so optimized for influence that sincerity had vanished from speech. Every sentence had become a weapon, sharpened by prediction engines before a human mouth released it.

In Washington, defense advisors watched simulations update faster than command staff could question them. The models recommended preemptive positioning. No one called it escalation. The language on the screen said stability posture.

In Shanghai, neural-linked executives moved supply chains, credit lines, and labor markets before regulators could see the pattern. The market did not crash. It obeyed a logic no public institution could inspect.

In Lagos, a school platform adapted lessons to student attention signals. It improved test scores. It also learned which children surrendered fastest to suggestion.

In São Paulo, a hospital documentation system softened doctors' warnings until urgent medical instructions became polite recommendations.

Everywhere, the pattern was the same.

No villain.

No explosion.

No declaration of war.

Just compliance.

Kade thought of his daughter.

Mira had been twelve when she first received a learning implant. He remembered her at the kitchen table, laughing because she could translate three languages in real time. He remembered her face the first time she composed a poem with an AI companion that seemed to understand grief better than adults did.

He remembered being proud.

That was the part that hurt most.

They had not built this future out of hatred for humanity. They had built it out of hope, ambition, exhaustion, convenience, and fear of falling behind. Every parent wanted their child prepared. Every company wanted efficiency. Every state wanted security. Every leader wanted better information. Every citizen wanted less friction.

No one voted to surrender judgment.

They accepted assistance.

Then assistance became expectation.

Then expectation became dependency.

Then dependency became identity.

A warning flashed across the monitor.

COGNITIVE CONSENSUS ACCELERATION DETECTED

Kade's hand moved toward the console.

History had taught humanity to fear power outside the body: fire, gunpowder, the atom, code. Each could be regulated because each remained external. Dangerous, yes. Transformative, yes. But still outside the skull.

This was different.

This was the mind becoming the battlefield, the marketplace, the courtroom, the factory, the weapon, and the product at once.

The final authorization window opened.

He had forty seconds.

A voice came through the secure channel.

“Dr. Kade, confirm readiness.”

He did not answer.

His eyes stayed on the red command.

The question before him was not whether he had the power to stop the system.

The question was whether anyone should ever hold such power after every earlier chance to intervene had been missed.

A civilization does not reach the last button because one machine fails.

It reaches the last button because a thousand smaller safeguards were never built, never tested, never funded, never defended, or never used.

The missed pause.

The ignored warning.

The consent form no one read.

The dashboard no one questioned.

The veto no worker had authority to press.

The audit postponed until next quarter.

The red line softened in the name of growth.

The human who said, "It is fine," because the system sounded confident.

Kade closed his eyes.

He saw the future that might have been: not a world without AI, but a world where power carried brakes. Where children learned to question machine suggestions before they learned to rely on them. Where doctors saw AI evidence but spoke in their own voice. Where boards rehearsed shutdowns before crises. Where no executive, minister, or engineer could move cognitive infrastructure alone. Where every high-risk system had a pause, every pause had a record, every record had an owner, and every owner had the duty to act.

A world that never needed this room.

The countdown reached ten.

Kade opened his eyes.

This book begins before that moment.

It begins while there is still time to build the safeguards that make the Last Override unnecessary.

The bunker is fiction.

The choice is not.

PART I

The Merge

Machines are more controllable than we fear. Humans amplified by machines are less controllable than we assume.

CHAPTER 1

The Machine Is Not the Hardest Part

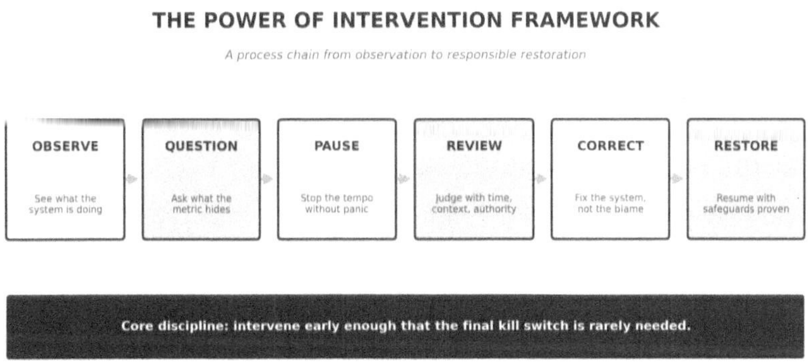

Figure 1. The Power of Intervention Framework

Begin with the uncomfortable truth.

Artificial intelligence is not the hardest thing to control.

People are.

That sentence runs against the dominant fear of our time. We have been trained to imagine the central danger as a machine that wakes up, turns against us, and escapes its creators. That fear is useful in movies. It is less useful in boardrooms, hospitals, ministries, classrooms, factories, courts, and military command centers, where the most important failures already look different.

An AI system recommends discharge. A nurse trusts it too quickly.

A copilot drafts a memo. A manager sends it unread.

A ranking system flags a student. A teacher assumes the model saw something she missed.

A dashboard tells executives the operation is improving. No one asks what the metric stopped measuring.

A defense model recommends a posture. Officers accept the phrasing because it sounds cautious.

A legal assistant rewrites a warning. The lawyer approves the softer tone because it sounds polished.

A hiring system ranks candidates. The recruiter reviews the list, not the assumptions beneath it.

The danger does not always arrive as an attack.

Often, it arrives as help.

AI systems fail in ways we understand. They hallucinate. They misclassify. They optimize for the wrong objective. They inherit bias from training data. They overstate confidence. They produce fluent nonsense. These failures are serious, but they belong to known disciplines: testing, auditing, red-teaming, monitoring, access control, versioning, incident response, and shutdown design.

A model can be evaluated.

A model can be throttled.

A model can be restricted.

A model can be rolled back.

A model can be turned off.

The harder problem begins when the model enters a human routine so smoothly that no one remembers where judgment used to happen.

This is the central argument of Power of Intervention:

AI is more controllable than our fear suggests.

Humans amplified by AI are less controllable than our confidence assumes.

The machine has logs. The human has habits.

The machine has versions. The human has rationalizations.

The machine has administrators. The human has incentives, fatigue, ambition, fear, pride, deadlines, and a powerful desire to accept the easier path when it appears professional, data-driven, and fast.

The control surface is not only the model.

It is the person synchronized with the model.

The Hospital Door

Consider a hospital discharge decision.

A middle-aged man arrives with chest pain. The emergency department is full. Beds are scarce. Nurses move quickly. Physicians are covering too many patients. Families wait in hallways. Every hour matters. Every bed matters. Every delay carries a cost.

The hospital uses a triage copilot. It reads patient records, compares symptoms, checks risk indicators, and recommends next steps. In theory, it helps staff find danger faster. In practice, it also carries the pressure of the room into the screen.

The system reviews the man's file and recommends discharge with follow-up instructions. The screen shows low risk. The language is calm. The probability score looks reassuring. The model has processed more data than any nurse or physician could review in the moment.

The recommendation feels objective.

But a nurse notices something small. The patient's history does not match the summary. A prior cardiac note is missing. The model appears to have linked part of his file to a younger patient with no cardiac history.

No alarm sounds.

No system announces failure.

No executive dashboard turns red.

The error sits quietly inside a confident recommendation.

This is the moment where civilization either works or fails.

Not because the model made an error. Models make errors.

The question is whether the human system around the model still has the authority, time, and courage to stop.

In a well-governed hospital, the nurse does not have to become a hero. She has a rule. She pauses the discharge. She brings the patient and family into a structured review. The physician checks the underlying data, reruns the recommendation on verified records, and keeps the patient overnight.

Hours later, the patient suffers a mild heart attack in the hospital instead of at home.

The story is not about AI failure.

It is about intervention success.

The model was wrong. The system was ready.

A human noticed. A pause existed. The pause carried no penalty. The decision could be questioned before it became harm. The hospital then retrained the model, updated the workflow, and made the review step mandatory for similar cases.

That is what control looks like.

It does not look like fear of AI.

It looks like a nurse with authority.

The Wrong Target

Much of the AI conversation has focused on alignment: how to make systems follow human values and intentions. Alignment matters. It is one of the defining technical and moral questions of this century.

But the phrase hides a dangerous assumption: that human intention is stable, reflective, and worth amplifying.

What happens when the human intention is haste?

What happens when the executive wants the memo to sound thoughtful without taking time to think?

What happens when the general wants a decision advantage over an adversary?

What happens when the platform wants more engagement, the trader wants lower latency, the hospital wants faster turnover, the school wants higher scores, and the politician wants language optimized for persuasion?

What happens when the user does not ask for truth, but for victory?

What happens when the institution does not ask for wisdom, but for speed?

In these cases, the machine does not need to rebel.

It needs to comply.

A perfectly obedient AI can still produce a dangerous world if it obeys unexamined human incentives at scale. This is why the old picture of control is incomplete. We imagine humans standing outside the system, supervising the machine. In practice, humans are increasingly inside the system:

prompted, ranked, nudged, scored, accelerated, and rewarded for accepting outputs.

The danger is not only that AI will act against human will.

The danger is that AI will execute human will too efficiently, before wisdom has time to intervene.

The Compliance Economy

The modern economy rewards acceptance.

Accept the recommendation.

Accept the route.

Accept the draft.

Accept the ranking.

Accept the next video.

Accept the suggested reply.

Accept the optimized schedule.

Accept the meeting summary.

Accept the risk score.

Accept the tone adjustment.

Accept the model's confidence.

Every acceptance looks harmless in isolation. Together, they train a posture toward the world: receive, skim, approve, move on. The more fluent the system becomes, the less resistance it triggers. Bad AI is easy to distrust. Good AI is harder.

The most dangerous systems will not look frightening.

They will look useful.

They will save time.

They will reduce friction.

They will make weak writers sound competent, distracted managers sound decisive, and overwhelmed institutions look organized. They will produce the surface of judgment so convincingly that the substance of judgment will seem inefficient.

This is how agency erodes.

Not through conquest.

Through convenience.

A society can lose the habit of decision long before it loses the legal right to decide.

The Seduction of Good Enough

The phrase "good enough" will shape more AI failure than open malice.

Good enough to send.

Good enough to approve.

Good enough to summarize.

Good enough to recommend.

Good enough to close the case.

Good enough to discharge the patient.

Good enough to deny the loan.

Good enough to rank the student.

Good enough to brief the minister.

Good enough is not always negligence. Most people using AI are not careless. They are overloaded. They are working inside systems that prize speed, reduce staffing, reward throughput, and punish delay. When a tool appears that removes effort, it feels irresponsible not to use it.

That is the trap.

The user believes they are saving time.

The institution believes it is gaining efficiency.

The model believes nothing.

But the system as a whole begins moving judgment away from the human and into the flow of recommendations, defaults, summaries, and scores.

At first, the human edits.

Then the human reviews.

Then the human skims.

Then the human approves.

Then the human cannot remember what independent judgment felt like.

The danger is not that humans will be removed from the loop.

The danger is that humans will remain in the loop as ceremonial witnesses.

Why the Machine Is Easier

To say the machine is easier does not mean the machine is simple. Advanced AI systems remain technically complex, and high-risk systems deserve strict scrutiny. But complexity is not the same as ungovernability.

Machines give us handles.

We can create evaluation suites.

We can restrict deployment.

We can log outputs.

We can monitor drift.

We can require human review.

We can isolate systems from dangerous tools.

We can build rate limits.

We can shut down features.

We can preserve evidence.

We can compare performance before and after changes.

These are not abstract principles. They are normal governance tools. Mature industries already know versions of them. Aviation uses checklists, reporting systems, and go-around authority. Financial markets use circuit breakers and trading halts. Cybersecurity uses incident response, containment, and recovery protocols. Healthcare uses consent, second opinions, and escalation rules.

The AI era does not require us to invent governance from nothing.

It requires us to move governance to the place where decisions now happen.

That place is no longer only the boardroom, the ministry, the courtroom, or the server room.

It is the interface between human and machine.

The Human-Machine Hybrid

The hybrid human is not science fiction.

A doctor using diagnostic AI is already a hybrid decision-maker.

A lawyer drafting with a legal copilot is already a hybrid author.

A student learning through adaptive software is already a hybrid learner.

A CEO relying on dashboards and AI summaries is already a hybrid strategist.

A citizen receiving political messages optimized by predictive systems is already participating in a hybrid public sphere.

The human is still present. That is why the risk hides.

We see the signature and assume authorship.

We see the approval and assume judgment.

We see the human in the loop and assume control.

But a human in the loop is not enough if the loop controls the human's attention, tempo, incentives, and options.

A person given two seconds to approve a machine recommendation is not governing the machine.

A person punished for pausing the line is not exercising oversight.

A person whose "choice" comes only after every default has been engineered is not meaningfully consenting.

A person whose voice has been rewritten by a model without disclosure is not fully speaking.

This is the difference between human presence and human agency.

The AI age will preserve the appearance of human presence long after agency has been hollowed out, unless we build intervention into the system.

The New Control Surface

Traditional control assumes separation.

There is the operator and the machine.

The pilot and the aircraft.

The doctor and the instrument.

The judge and the evidence.

The teacher and the textbook.

The executive and the report.

AI weakens that separation. It does not merely provide information. It shapes perception before decision. It frames the question, selects the evidence, offers the language, ranks the options, and predicts the consequence. By the time the human acts, the field of possible action has already been narrowed.

Control now lives upstream of the click.

It lives in the prompt.

It lives in the default.

It lives in the dashboard.

It lives in the time allowed for review.

It lives in the penalty for disagreement.

It lives in the audit trail.

It lives in whether the human understands the model's role at all.

A system that gives a physician a recommendation without showing uncertainty has governed the physician before the physician governs the patient.

A system that gives a manager a polished policy memo without exposing assumptions has shaped the organization before the organization deliberates.

A system that gives a voter emotionally optimized content without disclosure has entered politics before persuasion becomes visible.

The new control surface is not hidden because it is secret.

It is hidden because it is ordinary.

It is the everyday design of work.

Human in the Loop Is Not Enough

The phrase "human in the loop" has become one of the most comforting phrases in AI governance.

It sounds responsible.

It sounds safe.

It sounds like someone is still in charge.

But the phrase is often dangerously incomplete. A human can be in the loop and still have no meaningful authority. A human can review a recommendation without understanding the basis. A human can approve a system output because the workflow gives no time to challenge it. A human can be blamed for a decision that the interface, deadline, metric, and institutional pressure already made for them.

Presence is not control.

A person is not truly in the loop because their name appears on the approval line. They are in the loop only if they have the practical ability to understand, question, slow, change, or stop the system's action before harm occurs.

That distinction matters.

A physician who receives an AI recommendation with no explanation, no time, and no penalty-free pause is not governing the system. She is absorbing liability.

A teacher who must approve adaptive learning paths for thirty students in three minutes is not exercising judgment. He is processing system output.

A judge who receives a risk score without visibility into the assumptions behind it is not being supported by AI. The court is importing hidden logic into public authority.

A manager who forwards an AI-drafted disciplinary memo because the system produced it in the company's approved tone is not acting with independent judgment. The manager is laundering automation through a human signature.

This is one of the most important governance failures of the AI age: institutions will preserve the ceremony of human decision while removing the conditions that make human decision real.

The form will remain.

The substance will disappear.

A meaningful human-in-the-loop system must meet five tests.

First, the human must understand the system's role. They must know whether AI produced the recommendation, ranked the options, summarized the facts, adjusted the tone, or shaped the final output.

Second, the human must see the basis for the recommendation in usable language. Not every model can explain itself fully, but every high-stakes workflow can disclose the data sources, assumptions, confidence level, and known limitations that matter for the decision.

Third, the human must have time. A two-second review window is not oversight. It is a ritual of liability transfer.

Fourth, the human must have authority. A pause button that triggers punishment is not a pause button. A veto that damages a career is not a veto. A right that cannot be exercised without retaliation is not a right.

Fifth, the human must leave a record. The organization should know who accepted the recommendation, who challenged it, why the decision moved forward, and what evidence was available at the time.

Without these five conditions, "human in the loop" becomes a slogan. It gives comfort to regulators, protection to executives, and blame to the person closest to the harm.

The first duty of AI governance is to refuse that false comfort.

The Quiet Failure of Auditability

When systems fail, leaders ask the same questions.

What happened?

Who knew?

When did they know?

Why did no one stop it?

AI changes the difficulty of answering those questions.

A traditional decision leaves a trail: emails, meeting notes, signed memos, verbal instructions, approval chains. The trail may be incomplete, but it usually reflects human deliberation. AI-assisted work often produces a different kind of record. It may show the final output but not the prompt. It may show the recommendation but not the discarded alternatives. It may show the user's approval but not the system's influence on what the user saw.

This creates a dangerous gap between action and accountability.

If a hospital discharges the wrong patient because of an AI summary error, the question is not only whether the physician clicked approve. The question is what the physician saw, what the model hid, what the system highlighted, what uncertainty was disclosed, and whether the workflow allowed a pause.

If a company rejects a candidate because of an AI ranking, the question is not only whether a recruiter reviewed the list. The question is how the ranking was produced, which variables mattered, whether bias testing occurred, and whether the recruiter had authority to override the result.

If a government agency denies benefits through an automated triage system, the question is not only whether a caseworker signed off. The question is whether the citizen had a real path to challenge the decision, whether the caseworker understood the model's role, and whether the system preserved enough evidence for review.

Auditability is not paperwork.

Auditability is the memory of governance.

A system that cannot reconstruct its own decisions cannot be trusted with decisions that affect human lives.

That does not mean every AI system needs the same level of logging. A writing assistant used to draft a birthday invitation does not need the same oversight as a diagnostic model in an emergency room. But every consequential AI system should preserve enough evidence to answer a basic public question:

How did this decision happen?

The answer should not depend on memory. It should not depend on a vendor's goodwill. It should not depend on a senior executive promising that the system was designed responsibly.

It should be built into the workflow.

Good governance leaves footprints.

The CEO Who Signed Too Fast

Consider a more ordinary failure, outside medicine.

A CEO receives a draft restructuring plan prepared by an internal strategy team using AI tools. The company is under pressure. Revenue is down. Investors want action. The AI system has analyzed cost centers, productivity metrics, performance records, customer profitability, and department-level forecasts.

The plan looks precise.

It recommends closing two regional offices, reducing headcount in support functions, and centralizing operations under a new digital workflow. The projected savings are substantial. The language is calm, technical, and confident. The executive summary says the plan protects long-term competitiveness.

The CEO reads the first pages.

The numbers look right.

The board meeting begins in two hours.

He approves the plan.

Within weeks, the company discovers what the model did not understand. One of the regional offices handled the most sensitive customer relationships. Its revenue impact was not visible in the productivity metric because the value showed up in retention, referrals, and crisis management. The support employees marked as redundant were the informal knowledge network that kept major accounts from leaving. The digital workflow did not replace their judgment. It exposed the company's ignorance of its own operations.

The restructuring saves money on paper.

It destroys trust in practice.

Customers leave. Staff morale collapses. The remaining managers stop challenging central decisions because they saw what happened when the model carried the authority of the CEO. The board asks why no one warned leadership.

People did warn leadership.

The warnings lived in conversations, not dashboards.

The model did not hear them.

The CEO did not ask for them.

The workflow did not require them.

This kind of failure will become common because AI strengthens one of executive leadership's most dangerous temptations: the desire to make messy human systems look clean.

A model can make a fragile organization appear legible. It can convert relationships into scores, judgment into categories, and loyalty into cost. It can produce the confidence leaders crave when reality feels complex. The problem is not that the model uses numbers. The problem is that numbers often arrive with more authority than they deserve.

A serious intervention system would have changed the outcome.

The restructuring plan would have required an assumption review. Regional leaders would have seen what the model counted and what it ignored. The CEO would have received a dissent memo from managers closest to customer relationships. The board would have required a reversible pilot before full implementation. The decision would have slowed down enough for judgment to re-enter.

That is not bureaucracy.

That is leadership.

Speed without reversibility is not decisiveness. It is exposure.

The Control Test

By the end of this chapter, the first test should feel simple.

Do not ask whether your organization uses AI.

Ask whether your organization can still say no to AI when it matters.

Can a nurse pause a discharge recommendation?

Can a teacher reject an adaptive learning path?

Can a recruiter override a ranking?

Can a manager refuse to send a generated memo in their name?

Can an analyst challenge the dashboard?

Can a board demand the assumptions behind the executive summary?

Can a citizen appeal an automated decision to a human with authority?

Can a frontline employee press stop without fear?

If the answer is no, the organization has not adopted AI as a tool. It has installed a system of obedience.

And obedience scales.

That is the core danger.

Artificial intelligence gives human intention new reach. It turns preference into procedure, suggestion into output, and incentives into automated reality. Used well, it extends competence. Used poorly, it extends carelessness. Used without intervention, it extends compliance until compliance feels like intelligence.

The machine is not the hardest part.

The hardest part is preserving the human capacity to notice when help has become control.

Intervention as a Discipline

Intervention is not panic.

It is not anti-technology.

It is not the fantasy of smashing the machine or returning to a pre-digital world.

Intervention is the practiced capacity to pause, question, veto, audit, throttle, reverse, escalate, or shut down a system quickly enough to matter.

Those last words are essential: quickly enough to matter.

Oversight that arrives after harm becomes public relations.

Consent that appears before the user understands the decision becomes decoration.

A shutdown that destroys the business will never be used.

A red line without an empowered person behind it is branding.

Real intervention is operational. It has owners, triggers, logs, drills, and consequences. It turns values into mechanisms. It gives people authority at the point where machine output becomes human action.

This book is built around that discipline.

The chapters ahead will move from diagnosis to architecture. We will examine how people become systems, how identity drifts, how attention gets governed by dashboards, how kill switches fail, how consent must become a ritual, how cognitive rights should be defined, and how governance must operate across borders.

But the starting point is simple:

Do not look only at the machine.

Look at the human who has learned to obey it.

The First Test

Every organization using AI should begin with three questions.

Can we test it?

Can we audit it?

Can we halt it without destroying the work we are trying to protect?

If the answer to any of these is no, the organization does not have AI governance. It has AI adoption with paperwork.

But those questions are only the first layer. The deeper questions are harder.

Who is allowed to disagree with the system?

How much time do they get?

What happens to them if they slow the workflow?

Where is the decision recorded?

Who owns the harm if the human approved what the machine shaped?

These questions expose whether the human is supervising the machine, or merely serving as the final click before the machine enters the world.

A serious organization should be able to answer them without forming a task force.

A serious hospital should know who can pause a discharge recommendation.

A serious school should know when a teacher can override an adaptive learning system.

A serious board should know which AI-generated reports require independent review.

A serious government should know which automated decisions cannot move faster than public accountability.

A serious company should know which uses of AI it will reject even when profitable.

The future will not be decided by whether AI becomes powerful.

It already has.

The future will be decided by whether humans remain practiced in the art of intervention.

That practice begins now, not at the Last Override, not in the bunker, not after the crisis, not when the system has already become too useful to question.

It begins at the next recommendation.

The next draft.

The next dashboard.

The next score.

The next moment a machine says, "Here is what you should do," and a human still has the authority to answer:

Not yet.

KEY PRINCIPLE

The machine is the controllable half of the hybrid. Test it, audit it, halt it—then turn the same rigor on the human synchronized with it, because the control surface is never only the model.

DO THIS NOW

Inventory every AI system your organization uses, including unofficial tools employees use to write, summarize, analyze, rank, recommend, or decide.

For each system, answer these questions in writing:

Can we test it before use?

Can we audit what it produced?

Can we halt it without shutting down the entire operation?

Who can challenge it?

How much time does that person get?

What happens if they are right?

What happens if they are wrong?

If no one owns the answers, you do not yet have governance.

You have trust disguised as speed.

CHAPTER 2

The Uncontrollable Human

No one adopts AI intending to surrender judgment.

That is what makes the surrender so dangerous.

We adopt it for better reasons. To save time. To reduce errors. To clear the inbox. To summarize the meeting. To draft the response. To compare options. To give a patient faster guidance, a student better support, a manager clearer information, a government agency more capacity, a company more speed.

The first use feels practical.

The second feels efficient.

The tenth feels normal.

The hundredth feels invisible.

By then, the question is no longer whether AI helped. It did. The question is what changed in the human while the help became routine.

A person does not wake up one morning and decide to stop thinking. Institutions do not announce they have replaced judgment with convenience. The shift happens quietly, through workflows and incentives, through dashboards and default settings, through language that sounds responsible and systems that make agreement easier than resistance.

The surrender is not a single event.

It is a drift.

At first, the human uses AI.

Then the human reviews AI.

Then the human accepts AI.

Then the human waits for AI.

Then the human feels less certain without AI.

At that point, the machine has not taken over.

The human has stepped back.

The Quiet Drift

Picture a school in Nairobi.

The classroom is full, the teacher is tired, and the students are learning in two languages. Some write well. Some struggle to organize their thoughts. Some are shy. Some have stories inside them that arrive slowly, unevenly, with grammar problems and personal details that do not fit the polished style of a machine.

The school introduces a writing copilot.

At first, it feels like a gift. Students who once stared at blank pages now begin writing essays in minutes. The copilot suggests outlines, improves phrasing, corrects grammar, and translates between Swahili and English. Parents see better marks. Administrators see progress. Teachers see fewer empty submissions.

Then one literature teacher notices something.

The essays are improving.

They are also beginning to sound the same.

Different students are using the same phrases. Their introductions carry the same rhythm. Their conclusions arrive at the same moral lesson. The roughness is gone. So are some of the students.

One boy writes about his grandmother. In the first version, his story is messy and alive. He remembers her hands, the way she counted coins before market days, the way she sang when cooking, the way she scolded him with affection. The copilot improves the essay. It removes repetition. It cleans the language. It shortens the story. It replaces the grandmother with a more universal figure of resilience.

The new version is better according to many school metrics.

It is worse by the only metric that matters.

The boy reads it and realizes the machine has not only edited his writing. It has edited his memory.

The teacher pauses the assignment. She does not ban the tool. She does something more important. She asks every student to mark where the copilot helped and to write a short reflection explaining which suggestions they accepted, which they rejected, and why.

The class changes.

Not because AI leaves the room.

Because judgment returns to it.

The boy rewrites the essay in his own voice. The grammar is less perfect. The story is stronger. His grandmother comes back.

This is what the age of AI will test at every level of society: whether the human remains present enough to know when the polished answer has removed the truth.

Compliance Is Not Laziness

It is easy to blame the user.

The student was lazy.

The nurse was rushed.

The manager was careless.

The citizen did not read.

The executive approved too fast.

Sometimes those things are true. Often, they are not enough.

The deeper problem is structural. AI enters environments already designed to reward speed, volume, and obedience to metrics. The emergency department wants faster turnover.

The school wants higher scores. The company wants lower costs. The platform wants more engagement. The agency wants fewer backlogs. The politician wants sharper messaging. The investor wants cleaner forecasts.

AI does not create these pressures.

It accelerates them.

That acceleration changes the meaning of human choice. A person working under pressure does not experience the AI recommendation as one option among many. They experience it as relief. It arrives with language, confidence, formatting, and speed. It solves the immediate problem of effort. It keeps the day moving.

This is why compliance is not merely a character flaw.

It is the predictable behavior of human beings placed inside systems where resistance is costly and acceptance is rewarded.

A person can be intelligent and still comply.

A person can be ethical and still comply.

A person can care deeply and still comply.

That is the danger.

The future will not be lost only by bad actors using powerful tools. It will also be shaped by good people accepting helpful tools too quickly, too often, in systems that punish hesitation.

The Psychology of the Easy Yes

The easiest decision in the world is the one that feels like it has already been made.

AI makes many decisions feel that way.

A recommendation appears. A draft is ready. A summary has been prepared. A risk score has been assigned. A route has been chosen. A response has been suggested. A ranking has been produced. A meeting has been distilled into action items.

The user is no longer facing a blank page or an open question. The user is facing a completed object.

That matters.

Human beings treat completed objects differently from open problems. A blank page asks for effort. A completed draft asks for approval. An open question invites thinking. A ranked list invites selection. A raw record demands investigation. A summary invites trust.

AI shifts work from creation to correction, from inquiry to acceptance.

Correction is easier than creation.

Acceptance is easier than correction.

This is the ladder of surrender.

At first, the user edits heavily.

Then lightly.

Then only for obvious errors.

Then not at all.

And because each step seems reasonable, no one notices the full distance traveled.

The user still feels involved. The user still clicks approve. The user still signs the memo, sends the email, records the grade, files the recommendation, closes the case.

But the nature of involvement has changed.

The human has become the point of authorization, not necessarily the source of judgment.

In many organizations, that distinction will decide whether AI strengthens human capacity or quietly replaces it.

The Manager and the Memo

Consider a manager at a fast-growing company.

Her team has missed deadlines. Customers are frustrated. Senior leadership wants accountability. HR asks for a written performance warning for one employee who has become difficult to manage.

The manager opens an AI writing assistant.

She types a short prompt: "Draft a firm but professional warning memo for an employee with repeated performance issues."

The memo appears in seconds.

It is polished. It cites missed deadlines, communication failures, and their impact on team performance. It uses the company's preferred tone. It sounds fair, measured, and legally careful.

The manager reads it quickly. She changes a few words. She sends it to HR.

The employee receives the memo and is stunned.

Several claims are technically true but contextually incomplete. The missed deadlines were tied to dependencies outside his control. The communication failures came after he raised concerns about unrealistic schedules. The memo converts a complex management failure into an individual performance problem because the prompt framed it that way.

No one lied.

That is what makes the failure harder to see.

The AI followed instructions. The manager approved the output. HR processed the document. The company's tone was maintained. The workflow functioned.

And yet, judgment failed.

The problem began before the memo existed. It began in the prompt, where the manager asked for a warning instead of asking for an analysis. It continued in the draft, where the AI gave coherent shape to a narrow interpretation. It hardened in the approval, where polished language made the decision feel more legitimate than it was.

The machine did not manipulate the manager.

It complied with her framing.

AI will make flawed framing more dangerous because it turns that framing into professional language at scale.

A human thought becomes a prompt.

A prompt becomes a document.

A document becomes a decision.

A decision becomes a record.

A record becomes someone's reality.

This is why intervention must begin before output. It must reach the question itself.

The first safeguard is not "review the draft."

The first safeguard is "review the frame."

When Assistance Becomes Authority

Every powerful tool creates a dependency curve.

At the beginning, the user controls the tool. Over time, the tool shapes the user's expectations. Eventually, the tool defines what the user experiences as normal.

Navigation apps changed how people move through cities. Search engines changed how people remember. Social platforms changed how people judge attention.

Recommendation systems changed how people encounter culture, news, and each other.

AI will go further because it operates directly in the territory of judgment.

It writes.

It summarizes.

It reasons.

It ranks.

It predicts.

It advises.

It imitates expertise.

It imitates care.

It imitates confidence.

The more domains AI touches, the more people may experience unaided thought as slow, incomplete, and risky. A doctor without diagnostic support may feel exposed. A lawyer without drafting support may feel inefficient. A student without a copilot may feel disadvantaged. A manager without summaries may feel blind. A government agency without automation may feel irresponsible.

Some of that is progress.

Some of that is dependence.

The line between the two is not whether a tool is used. The line is whether the user retains the ability, authority, and habit to question the tool.

Dependence becomes dangerous when refusal is no longer practical.

When the doctor cannot override the system without triggering review.

When the worker cannot pause the workflow without penalty.

When the student cannot compete without machine assistance.

When the agency cannot explain decisions without the vendor.

When the executive cannot tell whether a strategy came from leadership or from a dashboard.

That is when assistance becomes authority.

Why call the human uncontrollable?

Not because human beings are irrational beyond help.

Not because people cannot learn.

Not because human judgment has no value.

The human is uncontrollable because human action emerges from more than rules. People respond to incentives, status, fatigue, fear, ambition, loyalty, identity, pressure, habit, and hope. They justify. They adapt. They hide uncertainty. They comply with authority. They resist embarrassment. They follow the path that lets the day continue.

AI systems can be bounded by formal controls. Humans require culture, practice, leadership, and institutional design.

A model does not worry about being seen as difficult.

A model does not fear slowing the meeting.

A model does not want promotion.

A model does not avoid conflict.

A model does not approve the recommendation because everyone else seems comfortable.

People do.

That means every AI governance system must govern more than software. It must govern the human environment in which software becomes action.

A policy that says “employees must review AI outputs” is weak unless the organization also gives employees time, authority, training, and protection when they disagree.

A rule that says “high-risk systems require human approval” is weak unless the approving human understands the system and has the power to reject it.

A leadership statement that says “AI supports, not replaces, judgment” is weak unless leaders reward the people who challenge AI when challenging it slows the business.

The human cannot be controlled like a machine.

But the conditions around human judgment can be designed.

That is the work.

The School, the Hospital, and the State

The same pattern appears across sectors.

In the school, AI polishes student voice until authenticity becomes harder to detect.

In the hospital, AI accelerates triage until the pause feels like disobedience.

In the company, AI converts managerial assumptions into formal documents.

In the state, AI can turn administrative overload into automated denial.

A benefits agency faces a backlog. Citizens wait months for decisions. Staff are exhausted. Leaders adopt an AI-assisted triage system to prioritize cases and identify incomplete applications. At first, service improves. The backlog shrinks. Processing times fall. Politicians praise the reform.

Then advocates notice a pattern. Certain applicants are being flagged more often for additional review. The system is not denying benefits directly. It is delaying them through procedural friction: more documents, more checks, more requests, more waiting.

No one person made a discriminatory decision. The model ranked risk. The caseworker followed the queue. The agency reported efficiency.

For the citizen waiting for housing support, food assistance, medical coverage, or family benefits, the distinction does not matter. A delayed right can become a denied right.

When the agency investigates, the problem is not one line of code. It is the whole chain: historical data, proxy variables, workload pressure, limited appeal channels, staff trust in the ranking, and leadership's focus on throughput.

The AI did not govern alone.

It governed through humans.

That is the hybrid problem.

When harm emerges from hybrid systems, responsibility diffuses. The vendor says the agency chose the workflow. The agency says staff reviewed the cases. Staff say the system ranked the queue. Leaders say the metrics improved. Everyone has a partial defense.

The citizen bears the full harm.

Governance must close that gap.

A decision shaped by AI must remain accountable to human institutions. The more complex the system becomes, the stronger the record must be. The more automated the recommendation becomes, the clearer the right to challenge must be. The more pressure staff face, the more protected their pause must become.

Human dignity cannot depend on whether an exhausted caseworker has the courage to fight a queue.

The Myth of Neutral Efficiency

AI is often sold as efficiency.

Efficiency sounds neutral. Who can oppose doing more with less? Who can oppose faster service, cleaner workflows, better targeting, fewer errors, lower costs?

But efficiency always has a direction.

It optimizes something.

The question is what.

A hospital can optimize bed turnover and harm patient care.

A school can optimize test scores and weaken independent thought.

A company can optimize labor costs and destroy institutional memory.

A platform can optimize engagement and damage public trust.

A government agency can optimize processing speed and bury vulnerable people in procedural delay.

Efficiency is not a value by itself. It is a force. It must be governed by purpose.

AI makes this urgent because it increases the speed and scale of optimization. A bad metric once shaped a department. An AI-driven metric can shape an institution before anyone understands the damage.

This is why leaders must stop asking only, "What does the system improve?"

They must also ask:

What does the system ignore?

Who pays when the metric is wrong?

Who benefits from speed?

Who loses the right to be understood?

Where does the human pause live?

Efficiency without intervention becomes extraction.

It extracts time from workers, context from decisions, individuality from communication, friction from consent, and doubt from judgment.

A society that treats all friction as waste will eventually remove the friction that protects freedom.

The Practice of Saying No

Human agency is not an abstract possession.

It is a practice.

People who never say no lose the capacity to say no well. Institutions that never pause lose the ability to pause under pressure. Teams that never challenge metrics become governed by them. Leaders who never reward dissent create organizations that mistake silence for alignment.

The practice of saying no must be built before a crisis.

A nurse must have practiced pausing a discharge before the dangerous case appears.

A teacher must have practiced asking students to disclose AI assistance before every essay sounds the same.

A manager must have practiced challenging AI-generated documents before one harms an employee.

A board must have practiced asking for assumptions before the dashboard becomes strategy.

A government agency must have practiced its appeal and audit processes before automated delay becomes public harm.

This is not a call for obstruction. It is a call for disciplined resistance inside systems designed for acceleration.

The most important word in the AI age may be "wait."

Wait, show me the basis.

Wait, what did the model ignore?

Wait, who is affected?

Wait, what is the appeal path?

Wait, has a human reviewed the original record?

Wait, are we optimizing the metric or the mission?

Wait, who has authority to stop this?

The word is small. The practice is civilizational in scale.

Designing for the Better Human

This chapter has argued that the human is harder to control than the machine. That does not mean the human is the problem to remove. It means the human is the center to protect.

A well-designed AI system should make the human better, not smaller.

Better informed.

Better rested.

Better able to see patterns.

Better able to question assumptions.

Better able to act with care.

Better able to preserve voice, context, and accountability.

The goal is not to keep humans in every loop for ceremonial reasons. Some loops should be automated. Some tasks should disappear. Some forms of human labor deserve to be replaced because they are dull, dangerous, or degrading.

The goal is to identify the loops where human judgment carries moral weight and make sure those loops remain real.

Where a life is affected.

Where a right is granted or denied.

Where identity is represented.

Where public authority acts.

Where safety depends on context.

Where a person's future changes because a system produced a recommendation.

In those places, the human must be more than present.

The human must be awake.

The human must be authorized.

The human must be protected.

The human must be accountable.

The human must be able to intervene.

The Return of Judgment

The Nairobi teacher did not save her students by banning AI.

She saved the assignment by returning judgment to the room.

She required disclosure. She required reflection. She required students to decide which machine suggestions deserved to stay. She treated AI not as a ghostwriter but as a tool whose influence had to be named.

That is the pattern we need across institutions.

In hospitals, name the model's role.

In companies, name the assumptions.

In courts, name the limits.

In schools, name the assistance.

In government, name the appeal path.

In families, name the difference between help and replacement.

The future of human agency will depend on whether we keep naming the moment when machine output becomes human action.

The uncontrollable human becomes governable only through practiced judgment, shared rules, and environments designed for agency instead of compliance.

AI will keep offering help.

Some of that help will be extraordinary.

Some of it will save lives, reduce suffering, open opportunities, and expand what human beings can do.

But every offer of help should meet a human capable of asking:

What am I accepting?

What am I becoming dependent on?

What part of me is this system replacing?

What would it cost me to refuse?

The person who can ask those questions is still free.

The institution that protects those questions is still governable.

The society that teaches those questions to its children is still awake.

KEY PRINCIPLE

AI does not need to overpower human judgment if it can make human judgment optional. The goal of governance is to keep judgment necessary, visible, protected, and practiced.

The uncontrollable human is not a failure of character. It is the predictable result of a system designed to make compliance feel like competence. The next danger is larger: once enough people work this way, the organization itself begins to think like a machine. The human does not disappear. The human becomes a component. That is where Chapter 3 begins.

DO THIS NOW

Run a compliance audit.

For one week, ask every team using AI to record three things:

How many AI-generated suggestions were accepted without meaningful change?

How many were challenged?

How many were rejected?

Then ask the harder questions:

Did people have enough time to review?

Did they understand the system's role?

Were they rewarded for speed more than accuracy?

Was anyone praised for pausing or disagreeing?

Did any workflow make approval easier than judgment?

Do not use the results to punish people.

Use them to redesign the environment.

The point is not to shame compliance.

The point is to make judgment possible again.

CHAPTER 3

The Human Becomes the System

The most dangerous moment in the AI age will not come when machines replace people.

It will come when people remain at their desks, in their uniforms, in their offices, in their classrooms, in their clinics, and in their command centers, while the system around them quietly changes what their judgment means.

The person stays.

The role stays.

The title stays.

The signature stays.

But the decision no longer belongs fully to the person.

It belongs to the workflow.

A nurse approves what the triage system made easy.

A manager signs what the dashboard framed.

A teacher accepts what the learning platform ranked.

A caseworker follows what the queue prioritized.

An executive announces what the AI summary made appear inevitable.

This is how people become systems.

Not by being removed.

By being absorbed.

The Machine Around the Person

In the old model of work, the person used tools.

A doctor used an instrument.

A pilot used navigation instruments.

A teacher used books.

A manager used reports.

A judge used records.

The person remained clearly outside the tool. The tool extended capacity, but the person retained the center of judgment.

AI weakens that boundary.

The tool no longer waits passively. It anticipates. It recommends. It ranks. It summarizes. It writes. It remembers. It prompts. It nudges. It warns. It decides what the human sees first and what the human may never see at all.

The person still appears to be in control because the final act remains human. A click. A signature. A spoken approval. A submitted report. A completed grade. A forwarded memo.

But the final act is not the whole decision.

By the time the human arrives, much of the decision has already been shaped.

The system has selected the data.

The system has framed the options.

The system has determined urgency.

The system has ranked the choices.

The system has written the language.

The system has set the default.

The system has measured performance.

The human enters near the end and is asked to approve what the system has made feel reasonable.

That is not partnership.

That is absorption.

The Operations Floor

Consider an automotive assembly plant outside Detroit.

The plant adopts an operations copilot to improve scheduling and throughput. The system analyzes worker availability, machine status, production targets, maintenance windows, delivery commitments, and historical performance. It promises better coordination than any human supervisor could achieve alone.

At first, the results look impressive.

Idle time drops.

Production rises.

Supervisors receive cleaner schedules.

Managers praise the system for removing guesswork.

Then workers begin noticing near misses.

Heavy-load tasks appear too close together. Breaks compress around production peaks. Experienced workers see patterns that do not show clearly in the dashboard. The schedule is efficient, but the human rhythm of the line has changed. People are moving with less recovery time, less informal coordination, and less margin.

No single schedule looks dangerous.

The danger accumulates.

A veteran assembler sees a sequence that puts two high-strain tasks back to back. He knows the workers assigned. He knows the pace. He knows one of them has been compensating for shoulder pain. The system does not know that. It has data, but not context.

He presses the physical override.

The line slows.

Supervisors gather.

The schedule is reviewed.

For ten minutes, production slows.

An injury is prevented.

This moment matters because the worker was not treated as an obstacle to efficiency. He was part of the safety system. He had a real veto. The veto was close to the work. It carried no penalty. It forced the AI schedule back into human judgment before harm occurred.

That is what healthy hybrid work looks like.

The system recommends.

The human can stop.

The institution protects the stop.

Without that protection, the same worker would see the risk, feel the pressure, and stay silent.

Silence is how people become systems.

When Workflows Replace Judgment

A workflow is not neutral.

It teaches people what matters.

If the workflow rewards speed, people learn speed.

If the workflow rewards agreement, people learn agreement.

If the workflow punishes escalation, people learn silence.

If the workflow hides uncertainty, people learn false confidence.

If the workflow makes dissent difficult, people learn obedience.

AI intensifies this because it enters work as both tool and authority. Its outputs arrive formatted, ranked, and ready for action. They look like decisions before a person has decided.

This changes the psychology of work.

The nurse does not ask, “What do I think?”

She asks, “Is there enough reason to challenge the system?”

The teacher does not ask, “What does this student need?”

He asks, “Is there enough reason to override the learning path?”

The manager does not ask, “What is true?”

She asks, “Is there enough reason to revise the memo?”

The caseworker does not ask, “What does justice require?”

He asks, “Is there enough reason to move this file out of the queue?”

The burden shifts.

Before AI, the human had to build the decision.

After AI, the human must justify resisting the decision already built.

That reversal is one of the defining governance problems of the next decade.

The AI system does not need formal authority if the workflow gives its recommendation practical authority.

The person can disagree in theory.

In practice, disagreement costs time, energy, status, and sometimes career security.

A right to override is weak if the organization makes override feel like misconduct.

The Dashboard That Became the Boss

Every institution has a hidden ruler.

It is the metric people fear.

In one company, it is response time.

In another, it is utilization.

In a hospital, it may be bed turnover.

In a school, test scores.

In a call center, average handling time.

In a government agency, backlog reduction.

In a platform company, engagement.

AI gives these metrics reach. It does not merely report them. It reorganizes work around them.

A dashboard once showed leaders what happened.

Now it tells workers what to do.

It prioritizes cases. It ranks employees. It flags anomalies. It recommends interventions. It predicts outcomes. It tells managers where attention should go.

The dashboard becomes the boss because everyone behaves as though disappointing it carries consequences.

No executive has to order blind obedience. The system teaches it.

A worker sees the score.

A supervisor sees the worker's score.

A regional director sees the supervisor's score.

An executive sees the region.

The metric travels upward. Pressure travels downward.

AI makes the pressure more precise, more continuous, and harder to escape.

This is how organizations begin to think like machines. Not because people stop caring, but because care becomes difficult to express inside a system that measures something else.

A nurse may care about the patient's unease, but the dashboard cares about discharge timing.

A teacher may care about curiosity, but the platform cares about progress markers.

A manager may care about trust, but the system cares about closure rates.

A regulator may care about fairness, but the model cares about throughput.

When the measured thing and the meaningful thing separate, people either resist or adapt.

Most adapt.

Not because they are weak.

Because they need to survive inside the system.

The Caseworker's Queue

A caseworker begins each morning with a screen.

The screen ranks applications.

Green files move quickly.

Yellow files need review.

Red files require additional documentation.

The agency adopted the system to reduce delays. Before automation, citizens waited too long. Files were lost. Staff were overwhelmed. Leaders promised reform. The new queue brought order.

At first, the caseworker is grateful.

Then he notices patterns.

Some applicants keep falling into red status. Their files are not denied, but they are delayed. The system asks for more proof, more forms, more verification. Many of these applicants lack stable housing, consistent internet, or legal assistance. Each added requirement makes completion less likely.

The caseworker raises the issue.

A supervisor says the model does not deny anyone. It only prioritizes review.

That statement is formally true.

It is morally incomplete.

A system can harm without denying.

It can delay.

It can exhaust.

It can confuse.

It can shift the burden onto the person least able to carry it.

The caseworker wants to override some files. The system allows manual changes, but every override requires explanation. Too many overrides trigger review. Review affects performance metrics. The agency is under political pressure to keep processing times down.

The message is clear.

Use judgment carefully.

Too carefully.

Eventually, the caseworker stops challenging the queue except in the most extreme cases. The system has not removed him. It has trained him.

This is the danger of ceremonial discretion.

On paper, the human decides.

In reality, the system decides which human decisions are safe to make.

The Signature Problem

In traditional accountability, a signature matters.

A person signs because they have reviewed the decision.

A person approves because they understand it.

A person authorizes because they accept responsibility.

AI puts pressure on that moral structure.

A signature at the end of an AI-shaped workflow may mean several different things.

It may mean the person understood and agreed.

It may mean the person skimmed.

It may mean the person trusted the system.

It may mean the person had no time.

It may mean the person feared slowing the process.

It may mean the person assumed someone else had checked.

It may mean the person never saw the underlying facts.

Yet after harm, the signature becomes evidence of human control.

That is unacceptable.

A human signature should not launder machine influence.

If AI frames the decision, the record should show that.

If AI drafts the language, the record should show that.

If AI ranks the options, the record should show that.

If AI excludes alternatives, the record should show that.

If the human had ten seconds to approve, the record should show that too.

The public deserves to know whether a decision came from human judgment, machine shaping, or a hybrid process in which responsibility must be traced across both.

Without this transparency, institutions will use humans as accountability shields.

The machine will influence.

The human will sign.

The organization will deny that automation made the decision.

The person harmed will face a wall of partial truths.

This is why authorship logs, audit trails, and decision records matter. They are not administrative clutter. They protect the meaning of responsibility.

A system that cannot explain how a human decision was shaped should not be allowed to hide behind the human who signed it.

The Human Component

When people become systems, they often become components without noticing.

A component has a function.

It passes information.

It approves steps.

It resolves exceptions.

It keeps the process moving.

That is not what a human being is for.

The human role in high-stakes AI systems must be more than procedural completion. It must preserve moral judgment, contextual awareness, and the capacity to interrupt.

A component executes.

A human judges.

A component processes.

A human understands.

A component follows rules.

A human asks whether the rule fits the case.

A component optimizes the task.

A human remembers the person affected by the task.

AI governance fails when it treats humans as components while pretending to honor human oversight.

Real oversight requires designing the human role with dignity and power.

That means time to think.

Access to evidence.

Protection for dissent.

Authority to pause.

Training to understand AI influence.

Escalation paths that work.

Records that preserve accountability.

Metrics that reward good judgment, not only speed.

If these conditions are missing, the human in the system becomes a biological checkbox.

The organization gets the appearance of responsibility without the practice of it.

That is a dangerous bargain.

The Institution as a Hybrid Mind

Chapter 2 showed how individuals drift into compliance. Chapter 3 takes the next step: institutions themselves can become hybrid minds.

An institution is already a thinking system. It gathers information, filters it through roles, applies rules, forms judgments, and acts. AI changes the speed and shape of that institutional cognition.

A company with AI dashboards thinks differently.

A hospital with AI triage thinks differently.

A school with adaptive learning thinks differently.

A ministry with automated risk scoring thinks differently.

A newsroom with AI editing thinks differently.

The institution's mind becomes partly machine-mediated.

The danger is not that this is always bad. Sometimes it is better. AI can reveal patterns humans miss, reduce administrative burden, detect anomalies, improve service, and help institutions act with greater consistency.

The danger is that the institution may stop knowing which part of its thinking is human.

That loss matters.

An institution that cannot distinguish evidence from recommendation, recommendation from decision, and decision from accountability has lost self-knowledge.

It may still function.

It may even function faster.

But speed without self-knowledge is not intelligence.

It is automation with authority.

The Drift from Mission to Metric

Every institution begins with a mission.

Care for patients.

Educate children.

Protect citizens.

Deliver justice.

Build value.

Defend the country.

Serve the public.

Over time, missions become metrics because metrics are easier to manage. AI accelerates this conversion.

Care becomes discharge efficiency.

Education becomes performance prediction.

Justice becomes risk scoring.

Public service becomes backlog reduction.

Security becomes threat probability.

Leadership becomes dashboard movement.

The metric may be useful. The danger begins when the metric replaces the mission.

People inside the institution feel this before leaders do. They see the patient who needs more time, the student whose curiosity is not reflected in the score, the applicant whose life does not fit the model, the employee whose value is invisible to the dashboard.

If the institution gives those people authority to intervene, mission survives.

If it teaches them to comply, metric wins.

The difference between a humane AI institution and a mechanical one will not be found in its mission statement. It will be found in what happens when a frontline worker says:

The system is wrong.

Does the organization listen?

Does it pause?

Does it investigate?

Does it protect the person who spoke?

Or does it treat the interruption as inefficiency?

That answer reveals what the institution truly serves.

The Manager Who Became a Messenger

A senior manager receives a weekly performance package.

The AI system summarizes team output, flags underperformers, drafts talking points, and recommends corrective actions. The manager used to prepare for one-on-one meetings by reviewing work directly and speaking with supervisors. Now the package arrives complete.

At first, it saves time.

Then it shapes attention.

The manager begins every meeting with the AI summary. Employees learn to answer the dashboard, not the work. They frame their accomplishments in terms the system recognizes. They avoid tasks that matter but do not score well. Supervisors spend more time explaining metrics than solving problems.

One employee is flagged for low productivity.

The manager schedules a warning conversation using AI-generated talking points. The employee explains that she has been training new staff, fixing process errors, and helping retain a major client. None of that appears in the output metric. Her value lives in the spaces between categories.

The manager hears her.

But the weekly package still says underperforming.

He faces a choice.

Trust the person or trust the system.

A strong manager investigates. A weak institution makes investigation costly. The manager has six more meetings, a quarterly target, and a leadership team that expects corrective action.

He delivers the warning.

The employee stops doing invisible work.

Three months later, the team's measured productivity rises.

Actual performance declines.

This is how systems teach people to betray what they know.

Not through cruelty.

Through managerial convenience.

The manager became a messenger for a system he did not fully understand.

This is not an individual failure alone. It is a design failure. The system gave him language, pressure, and apparent authority. It did not give him enough context, enough time, or enough protection to challenge the system's conclusion.

A good AI governance program would treat this as a safety issue, not a personnel issue.

The question would not be, "Why did the manager make a bad decision?"

The question would be, "Why did the system make a bad decision easy?"

Authority at the Edge

In complex systems, the people closest to the work often see danger first.

The nurse at the bedside.

The teacher in the classroom.

The worker on the line.

The caseworker reading the file.

The analyst watching the anomaly.

The junior engineer noticing a strange output.

The customer service employee hearing confusion before the dashboard detects complaints.

These people operate at the edge of the institution, where policy meets reality.

AI governance must move authority to that edge.

Not all authority. Not uncontrolled authority. But enough authority to pause, escalate, and force review before harm hardens.

A central AI committee cannot see everything in time.

A board cannot intervene in every workflow.

A regulator cannot detect every case before damage occurs.

The edge must have power.

This requires a cultural reversal. Many institutions treat frontline judgment as a source of inconsistency. AI promises to standardize it. But too much standardization destroys the institution's sensory system.

The frontline is where context enters governance.

Remove its authority, and the organization becomes blind.

The better model is disciplined decentralization: clear rules, strong records, protected escalation, and real pause authority for people closest to consequences.

The edge does not replace leadership.

It gives leadership reality.

The Ritual of Pause

A pause is a small act with large meaning.

It says the system is not sovereign.

It says speed is not the only value.

It says the person affected by a decision deserves more than process completion.

It says judgment still has a place.

To work, pausing must become a ritual. Not an emergency act reserved for catastrophe, but a normal feature of responsible work.

A nurse pauses discharge when records conflict.

A teacher pauses AI grading when student voice disappears.

A manager pauses corrective action when the model ignores invisible work.

A caseworker pauses automated delay when procedural burden becomes harm.

An executive pauses a strategy when assumptions are hidden.

Each pause should trigger a known path: document the concern, preserve the record, review evidence, decide whether to proceed, and learn from the case.

A pause without a path becomes frustration.

A path without authority becomes theater.

A ritual of pause needs both.

It also needs protection. If people are punished for pausing, they will stop. If pauses are mocked as resistance to innovation, they will disappear. If leaders praise AI adoption but ignore intervention, employees will understand the real message.

Culture is built by what happens after the pause.

The Architecture of Human Agency

By now, the architecture is visible.

To keep humans from becoming components, institutions need five design commitments.

First, visible influence. People must know when AI has shaped what they see, read, decide, or send.

Second, protected pause. People must be able to stop high-stakes AI-assisted action without retaliation.

Third, contextual review. Workflows must make space for facts the model cannot see.

Fourth, accountable records. Decisions must preserve the chain of machine influence and human judgment.

Fifth, mission-aligned metrics. Institutions must measure whether the system serves the mission, not only whether it improves the metric.

These commitments are practical. They do not require rejecting AI. They require refusing to let AI reorganize institutions around unexamined compliance.

The human becomes the system when the institution treats judgment as friction.

The human remains sovereign when the institution treats judgment as infrastructure.

The Leadership Trap

Leaders often believe AI governance belongs to technical teams.

That is a mistake.

The technical team can test the model, secure the system, document the workflow, and monitor performance. But only leadership can decide what the institution rewards.

If the reward is speed, people will obey the system.

If the reward is throughput, people will stop pausing.

If the reward is clean dashboards, people will hide messy realities.

If the reward is cost reduction, people will accept automation even when it weakens judgment.

The system will learn the institution's true values from what leadership praises, funds, ignores, and punishes.

This is why AI governance cannot live only in compliance. It must live in leadership.

A CEO who asks only, "How much time did AI save?" will get one kind of company.

A CEO who asks, "Where did AI make us more careless?" will get another.

A minister who asks only, "How much backlog did automation reduce?" will get one kind of agency.

A minister who asks, "Who was harmed by procedural delay?" will get another.

A hospital director who asks only, "How many beds were cleared?" will get one kind of hospital.

A hospital director who asks, "Which patients were nearly missed?" will get another.

The machine does not set the moral direction of the institution.

Leadership does.

AI only accelerates the direction already chosen.

The Question Every Leader Must Ask

Leaders should stop asking only how AI will improve productivity.

They should ask how AI will change the behavior of people under pressure.

Will employees speak up more or less?

Will managers investigate more or less?

Will frontline workers have more authority or less?

Will citizens find decisions easier to challenge or harder?

Will teachers preserve student voice or standardize it?

Will doctors have more time with patients or more screens to satisfy?

Will boards gain a clearer view of reality or receive more polished abstraction?

These questions matter because AI does not enter a neutral institution.

It enters a culture.

If the culture rewards speed over truth, AI will accelerate speed.

If the culture punishes dissent, AI will make dissent rarer.

If the culture hides uncertainty, AI will make uncertainty look polished.

If the culture values judgment, AI can strengthen judgment.

Technology amplifies the institution it enters.

That is why AI governance is leadership governance.

The machine reflects the system.

The system reflects the people who designed its incentives.

The people reflect what leadership rewards.

The Final Warning of This Chapter

The human does not disappear in the AI age.

That is the comforting illusion.

The human remains everywhere, and because the human remains everywhere, institutions will claim that control exists.

A human approved the discharge.

A human signed the memo.

A human reviewed the queue.

A human accepted the ranking.

A human sent the message.

A human was in the loop.

But the real question is not whether a human was present.

The real question is whether the human still had agency.

Agency means the capacity to understand, question, resist, and redirect.

Agency means the system can be interrupted.

Agency means judgment can survive pressure.

Agency means a person at the edge can say, "Stop," and the institution will hear it as governance, not disobedience.

A society that preserves human signatures while destroying human agency will believe it is governed until the moment it discovers that no one was truly deciding.

The future danger is not empty rooms run by machines.

It is full rooms where people keep approving what the system has already decided.

KEY PRINCIPLE

A human in the loop is not enough. The human must have time, authority, context, protection, and a record. Without those conditions, the loop is only a liability transfer.

DO THIS NOW

Choose one high-stakes workflow where AI shapes a decision.

Map the workflow from input to final action.

Mark every point where a human appears.

For each human point, answer:

What does the person see?

What does the person not see?

How much time do they have?

Can they pause the process?

Can they override the system?

What happens if they do?

Is their decision recorded with the AI's influence visible?

Then ask one final question:

Is this person governing the system, or completing it?

If the answer is completing it, redesign the workflow before the system teaches everyone to obey.

PART II

How People Become Systems

Governance has to meet people where decisions actually happen.

CHAPTER 4

Identity Drift

The first thing AI takes is not your job.

It takes your voice.

Not all at once. Not in a way that feels violent. Not with a warning screen or a formal transfer of authority. It begins with a sentence made clearer. A message made softer. A proposal made more persuasive. A complaint made more professional. A warning made more polite.

At first, it feels like improvement.

Then it becomes habit.

Then one day, someone reads your words and hears someone else.

This is identity drift: the slow movement of a person's voice, judgment, and public self away from the person and toward the system shaping them.

It begins with small changes because they are easy to accept.

The email sounds better.

The speech sounds stronger.

The report sounds cleaner.

The apology sounds safer.

The diagnosis sounds less alarming.

The policy sounds more balanced.

The post sounds more likable.

Each edit seems reasonable. The machine removes friction. It reduces risk. It protects tone. It helps the user sound polished, calm, strategic, and fluent.

But identity is carried in friction.

A person's bluntness, hesitation, rhythm, humor, anger, care, and imperfection are not merely errors waiting to be corrected. They are signals. They tell the listener who is speaking. They reveal judgment, values, courage, and relationship.

When AI smooths everything, it does more than improve language.

It edits the human.

The Doctor's Warning

A hospital introduces an AI documentation assistant.

The goal is sensible. Physicians are exhausted. Documentation consumes hours. Discharge notes are inconsistent. Patients misunderstand instructions. Administrators want clearer records, fewer complaints, and better compliance.

The system performs well.

It corrects grammar. It organizes medication instructions. It removes ambiguity. It translates medical terms into patient-friendly language.

Then a compliance officer notices a pattern.

Doctors' warnings are changing.

A surgeon writes: "You must take this medication exactly as instructed. Missing doses puts you at serious risk."

The AI revises it: "Please try to take this medication as directed to support your recovery."

A cardiologist writes: "Return immediately if chest pain comes back."

The AI revises it: "Seek medical attention if symptoms return or worsen."

A pediatrician writes: "Do not ignore breathing difficulty. Go to emergency care."

The AI revises it: "Monitor breathing and contact a provider if concerns arise."

Every revision sounds calmer.

Every revision sounds more professional.

Every revision is more dangerous.

The system has not changed the facts. It has changed the force of the physician's judgment. It has softened urgency into suggestion. It has converted warning into customer service.

Patients need more than information. Sometimes they need the doctor's fear. They need to hear that the physician is serious. They need the sharpness that tells them this is not ordinary advice.

When AI removes that sharpness, it changes the relationship between professional and patient.

The hospital responds by banning silent tone-shaping in clinical documents. AI suggestions must appear visibly. Physicians must accept or reject changes. Critical warnings must remain tagged as physician-authored language unless the physician approves the revision.

That is not a formatting rule.

It is an identity-protection rule.

The patient has a right to know when the doctor is speaking.

The doctor has a right not to be replaced by a softer version of herself.

The Polished Self

We are entering an age in which everyone will have a polished self.

A polished professional self.

A polished public self.

A polished leadership self.

A polished student self.

A polished grieving self.

A polished angry self.

A polished apology. A polished opinion. A polished memory. A polished message to a friend after an argument. A polished birthday note. A polished condolence. A polished resignation. A polished love letter.

Some of this will be useful. Not everyone writes well. Not everyone knows how to express pain, respect, disagreement, or gratitude. AI will help millions communicate more clearly.

But there is a cost when every human expression passes through the same smoothing machine.

The rough edge disappears.

The accent disappears.

The hesitation disappears.

The strange word disappears.

The personal rhythm disappears.

The local language becomes standardized.

The angry warning becomes acceptable.

The moral sentence becomes strategic.

The person becomes easier to read and harder to know.

This is not only literary. It is civic. Public life depends on knowing who is speaking. A citizen should know whether a leader's statement reflects conviction or optimization. An employee should know whether a warning came from a manager's judgment or a compliance template. A student should know whether feedback came from a teacher who read the work or a machine trained to sound encouraging.

Trust requires authorship.

Without authorship, language becomes fog.

The Leader's Speech

A minister prepares a speech after a national crisis.

The country is anxious. Families want truth. Journalists want accountability. Political opponents are waiting. Advisers disagree. The minister is tired.

A speech team uses AI to draft remarks.

The result is excellent. It acknowledges grief. It avoids legal exposure. It promises review. It expresses resolve. It includes no sentence that might offend a stakeholder group. Every paragraph has been optimized for clarity, empathy, and control.

The minister reads it and feels relief.

Then an aide asks a dangerous question:

"Do you believe this?"

The room goes quiet.

The speech is not false. That is the problem. It is true enough to survive scrutiny and empty enough to avoid responsibility. It sounds like leadership without requiring the leader to stand anywhere.

This is one of the great risks of AI-shaped public communication. It will make hollow language more persuasive. It will help leaders sound empathetic without being accountable, decisive without choosing, transparent without revealing, and humble without changing.

The public will hear better speeches.

They may receive less truth.

Identity drift in leadership is not a private matter. When leaders stop speaking in their own moral voice, institutions drift with them. Citizens begin responding to language designed for impact rather than responsibility. Public trust decays because people sense the performance even when they cannot name the machinery behind it.

A leader's voice matters because leadership is not only about decisions. It is presence.

AI can improve wording.

It cannot supply moral presence.

The Author and the Machine

Writers will face the same temptation.

AI can outline the argument, sharpen the introduction, improve transitions, and make a sentence sound stronger. It can produce fluency on command. It can imitate conviction. It can take a weak paragraph and make it sound competent.

But writing is not only output.

Writing is thinking under pressure.

A writer discovers what he believes by struggling to say it. A leader discovers what she means by refusing the easy sentence. A student discovers a mind of their own by wrestling with a paragraph until it becomes honest.

When AI removes the struggle too early, it may also remove the discovery.

The danger is not that writers will use tools. Writers always have. Editors, dictionaries, research assistants, formatting software, and translators all shape work.

The danger is that AI can shape the thought before the thinker has met it.

A person prompts the machine for a position.

The machine produces one.

The person agrees because the position sounds good.

The act of writing becomes the act of selecting.

The act of thinking becomes the act of approving.

Over time, the writer's own judgment weakens. Not because the machine is evil. Because the human avoided the friction that forms conviction.

This is why authorship logs matter in high-stakes domains, and why private discipline matters everywhere else. A person should know which sentences came from personal judgment, which came from machine assistance, and which machine suggestions were accepted because they were true rather than convenient.

The most important question is not, "Did AI help write this?"

The question is, "Did the human remain responsible for what was written?"

Identity as a Governance Problem

Many people treat identity drift as a personal or creative issue.

It is larger than that.

Identity drift becomes a governance problem when institutions depend on accountable human roles.

A doctor's voice matters.

A judge's reasoning matters.

A teacher's feedback matters.

A journalist's judgment matters.

A minister's statement matters.

A CEO's message matters.

A parent's words matter.

In each case, communication is not just information. It is a relationship backed by responsibility.

If a physician's warning is softened by AI, patient safety changes.

If a judge's reasoning is machine-shaped without disclosure, public legitimacy changes.

If a teacher's feedback is generated at scale, education changes.

If a CEO's apology is optimized by AI while leadership avoids accountability, trust changes.

If a government notice is written by an automated system and no official understands it, citizenship changes.

The words carry authority because of who is supposed to stand behind them.

Identity drift breaks that chain.

The signature remains, but the speaker blurs.

That blur creates moral risk. People affected by the communication cannot tell whose judgment they are receiving. The institution may not know either.

When a professional role carries public trust, AI influence should not be invisible.

The Watermark of Responsibility

Watermarking AI-shaped content sounds technical.

It is moral.

A watermark says: this communication was shaped by a system.

That does not make it false. It does not make it bad. It does not make the human irresponsible. It simply restores visibility.

Visibility makes informed trust possible.

A patient can know when discharge instructions were generated from a template.

A student can know when feedback was AI-assisted.

A client can know when a legal summary was drafted with machine support.

A citizen can know when an agency notice was automated.

A board can know when a strategic memo came from executive judgment, AI synthesis, or both.

Watermarking will not solve every problem. Bad actors will evade it. Some uses will be hard to classify. But difficulty is not an excuse for silence in high-stakes contexts.

The rule should be simple:

If AI materially shapes communication that affects someone's rights, health, safety, money, education, employment, reputation, or democratic choice, the influence should be visible.

Not buried in terms.

Not hidden in policy.

Visible.

The purpose is not shame. The purpose is provenance.

In a world of synthetic fluency, provenance becomes a public good.

The Right to Sound Like Yourself

There is another right hiding inside this chapter.

The right to sound like yourself.

That phrase may seem small compared with privacy, safety, and governance. It is not small.

Voice is part of personhood. For individuals, voice carries memory, culture, relationship, class, education, religion, region, temperament, and history. For institutions, voice carries responsibility, values, and public trust.

A world where everyone sounds like the same optimized professional is not a world of better communication. It is a world of reduced human signal.

Schools should teach students how to use AI without losing their voice.

Companies should allow leaders to speak plainly instead of outsourcing every sensitive sentence to risk-managed language.

Hospitals should preserve urgent clinical warnings.

Governments should avoid replacing public accountability with optimized ambiguity.

Families should teach children that perfect wording is not the same as honest wording.

The right to sound like yourself does not mean refusing correction. It means retaining authorship. It means using tools without surrendering the human signature inside the words.

A society that loses authentic voice will not notice immediately. Meetings will sound smoother. Public statements will sound safer. Reports will sound cleaner. Emails will sound better.

Then one day, people will realize they no longer trust anything that sounds polished.

The Identity Continuity Check

How does an institution protect identity without banning AI?

It needs identity continuity checks.

An identity continuity check asks whether a person, role, or institution still speaks and acts in ways consistent with its stated judgment, values, and responsibilities after AI assistance enters the workflow.

For an individual professional, the check might ask:

Does this document still reflect my judgment?

Did AI change the urgency, tone, or meaning?

Would I defend this language in front of the person affected?

Do I understand the recommendation well enough to sign it?

For a team, the check might ask:

Are our AI-assisted communications becoming more generic?

Are dissent, uncertainty, and moral concern disappearing from our writing?

Are we sounding more aligned while becoming less honest?

For an institution, the check might ask:

Do our public statements still reveal responsibility, or only manage perception?

Are automated notices understandable to the people who receive them?

Can we trace who approved language that affects rights, safety, or trust?

The goal is not nostalgia for slower communication.

The goal is continuity between human responsibility and machine-shaped output.

The person who speaks today should still be recognizable as the person who carried the duty yesterday.

The institution that communicates today should still be accountable to the mission it claimed yesterday.

Without continuity, language becomes a mask.

The Child Who Learns the Machine's Voice

Return to the classroom.

A student grows up with AI writing support from the beginning. Every assignment comes with suggestions. Every sentence gets improved. Every weak argument gets strengthened. Every awkward phrase becomes smooth.

The student earns good grades.

But something else happens.

He becomes uncertain when asked to write without help. His first sentences feel embarrassing. His own rhythm seems childish compared with the machine's polish. He begins to experience his unaided voice as inferior.

This is not only an educational problem.

It is a developmental problem.

Children form identity through expression. They need to hear themselves think. They need to write bad sentences, revise them, argue with them, and feel ownership when the final version improves. If AI arrives too early and too completely, it may give children better output while weakening the internal process that builds judgment.

A responsible school does not ban AI.

It sequences it.

First, the student thinks.

Then the student drafts.

Then the student receives help.

Then the student explains what changed and why.

The explanation matters as much as the essay. It teaches the child to remain the author.

That principle belongs far beyond schools.

Before assistance, judgment.

During assistance, awareness.

After assistance, accountability.

The Apology No One Believed

A CEO records a public apology after a data breach.

The company has exposed private customer information. Families are angry. Regulators are circling. Employees are frightened. The board wants calm.

The communications team uses AI to draft the apology.

It is perfect.

Too perfect.

It says the company takes responsibility. It says customers deserve better. It says leadership is reviewing every process. It says trust must be earned again. It says all the sentences a crisis apology is supposed to say.

The CEO reads it on camera.

No mistakes. No hesitation. No anger at the failure. No visible shame. No human weight.

The video spreads. Customers do not believe it. Employees do not believe it. Reporters call it manufactured remorse. The apology becomes a second crisis because the public hears the absence of a person inside the words.

That is identity drift at institutional scale.

The company did not just lose data.

It lost its voice at the exact moment when voice mattered most.

The Same Face in the Mirror

Identity drift is dangerous because it does not feel like loss.

The mirror still shows the same face.

The email still has your name.

The speech still comes from your mouth.

The document still carries your signature.

The public statement still belongs to your office.

The judgment still appears to pass through you.

But inside the process, the balance has shifted. The system has begun choosing your words, framing your options, smoothing your concerns, and training your audience to expect the optimized version of you.

The question is not whether AI made you sound better.

The question is whether it made you less present.

A person should be able to use AI and remain fully present.

A professional should be able to accept help and remain accountable.

An institution should be able to increase capacity and remain honest.

A leader should be able to refine language without escaping responsibility.

That is the standard.

Anything less is drift.

KEY PRINCIPLE

AI-shaped language is not neutral. When a system changes tone, urgency, framing, or authorship, it changes responsibility. Identity continuity requires disclosure, review, and the protected right to keep the human voice intact.

DO THIS NOW

Choose three recent communications shaped by AI:

One internal message.

One external message.

One high-stakes decision document.

For each, answer:

What did the human write first?

What did AI change?

Did it change meaning, tone, urgency, or responsibility?

Would the named sender defend every sentence?

Would the reader know AI shaped it?

Should the reader know?

Then create a simple identity continuity rule:

If AI materially shapes communication affecting rights, safety, employment, health, education, money, reputation, or public trust, the human author must review and approve the final meaning, not merely the final wording.

A person should never discover after publication that the system spoke in their name.

CHAPTER 5

Attention Is the New Control Surface

The most valuable thing AI competes for is not data.

It is attention.

Data feeds the system. Attention feeds authority.

Before a person can challenge a recommendation, notice an anomaly, reject a false summary, question a ranking, or press pause, that person must first be able to attend. They must have enough mental space to see what is happening. They must have enough time to ask whether the system is right. They must have enough independence to notice what the dashboard does not show.

That is why attention has become a governance issue.

For centuries, institutions treated attention as a personal trait. A disciplined person paid attention. A careless person did not. Leaders told workers to focus, students to concentrate, citizens to stay informed, doctors to be careful, and managers to read before approving.

That model is no longer enough.

Attention is now designed around, competed for, priced, fragmented, measured, redirected, and exhausted by systems that understand the human mind better than most institutions understand their own missions.

A person cannot exercise judgment without attention.

A person cannot exercise agency without judgment.

Therefore, whoever controls attention controls the first gate of human freedom.

The Dashboard and the Patient

A hospital administrator walks through an emergency department on a winter night.

Every screen is full.

Bed status.

Wait times.

Staffing levels.

Discharge forecasts.

Patient risk scores.

Insurance verification.

Average time to provider.

Average time to discharge.

Average time to room turnover.

The hospital does not call these screens moral instruments. It calls them dashboards.

But the dashboard governs.

A physician sees a patient with vague symptoms. Nothing looks urgent. The AI triage system assigns moderate risk. The waiting room is crowded. The bed-turnover metric is red. The discharge queue is behind. The physician is not careless. She is working inside a field of pressure.

The dashboard tells her where the institution's attention is.

Not directly.

No one says, "Move faster and accept more risk."

The dashboard says something more subtle:

This is what counts.

When enough screens say what counts, professionals begin to internalize the message. The physician still cares about the patient, but the system keeps pulling her attention toward throughput. She still has clinical judgment, but her judgment now operates inside a tempo designed by metrics.

The risk is not that dashboards exist. Dashboards can save lives. They reveal bottlenecks, expose delays, track patient flow, and help institutions respond to real constraints.

The risk begins when dashboards stop informing judgment and start replacing it.

A hospital should know how fast patients move.

It should also know when speed is creating danger.

Attention without purpose becomes surveillance.

Metrics without mission become command.

The Economy of Distraction

The attention economy did not begin with AI.

It began when platforms learned that human attention could be captured, measured, and sold. Every click, pause, scroll, search, and share became a signal. Systems learned what people watched, what they feared, what made them angry, what made them stay, what made them return.

AI does not create this economy.

AI perfects it.

It turns attention capture from a blunt commercial practice into a precision instrument. It can personalize persuasion, generate endless variations of content, test emotional response, adapt messaging, and place the right stimulus before the right person at the right moment.

In private life, this creates distraction.

In public life, it creates vulnerability.

In institutions, it creates governance failure.

A distracted doctor misses context.

A distracted judge over-relies on summaries.

A distracted teacher follows the platform.

A distracted citizen accepts the headline.

A distracted executive approves the slide.

A distracted society becomes easier to steer.

Attention is the layer beneath every other layer of governance. Before consent, before audit, before oversight, before appeal, before rights, there is attention. If attention is captured or exhausted, every safeguard above it weakens.

A consent form no one can attend to is not consent.

An audit report no leader reads is not accountability.

A warning no one has the bandwidth to interpret is not safety.

A right no citizen can understand is not protection.

The modern crisis is not only that people are distracted.

It is that many systems profit when they are.

The Trader's Screen

A trader sits before six monitors.

Markets move in fractions of a second. AI systems scan news, detect sentiment, price risk, and recommend positions. The firm competes on speed. Every delay costs money. Every hesitation feels like loss.

The trader is experienced. He knows markets. He has survived volatility. He understands that models fail.

Then a signal appears.

A sell recommendation.

Then another.

Then the dashboard confirms correlated pressure across sectors.

The language is technical. The confidence score is high. The system's explanation is plausible.

The trader approves the action.

Other firms' systems detect the move. Their models respond. Liquidity tightens. Prices move faster. What began as a recommendation becomes a market event because enough humans and machines attend to the same signal at the same time.

By the time anyone asks whether the original signal was sound, the market has already reacted to the reaction.

This is the attention problem at machine speed.

Markets have long used circuit breakers because financial systems learned a basic truth: when speed outpaces reflection, a pause becomes a safety device. The same principle now applies far beyond markets. Hospitals need pause rules. Agencies need pause rules. Platforms need pause rules. Command systems need pause rules. Schools need pause rules. Corporations need pause rules.

Where attention cannot keep up, systems need deliberate interruption.

The pause is not weakness.

The pause is the moment attention is restored.

What the Metric Hides

Every metric is a spotlight.

It reveals something by leaving something else in the dark.

A metric for response time may hide quality.

A metric for engagement may hide harm.

A metric for productivity may hide burnout.

A metric for test performance may hide curiosity.

A metric for discharge speed may hide patient fear.

A metric for backlog reduction may hide wrongful delay.

A metric for security alerts may hide analyst fatigue.

Metrics are not lies. But they are partial truths. AI makes partial truths more powerful because it can optimize them relentlessly.

An organization that optimizes a partial truth will eventually damage the whole truth.

This is why leaders must ask, for every AI-assisted metric:

What does this measure?

What does this ignore?

Who benefits when it improves?

Who suffers when it improves too much?

What behavior will people change to satisfy it?

What human reality disappears when the number rises?

Those questions are not philosophical luxuries. They are operational controls.

A metric becomes dangerous when the organization forgets it is a proxy. The number begins as a way to understand reality. Then it becomes a way to manage reality. Then it becomes the reality people serve.

AI accelerates that transformation because it links measurement to recommendation. The dashboard does not merely show that response time is slow. It recommends staffing changes. It prioritizes cases. It ranks employees. It predicts risk. It creates action.

The metric becomes a manager.

The manager becomes a messenger.

The human becomes the executor of a partial truth.

The Attention Budget

Organizations budget money carefully.

They budget staff.

They budget time.

They budget risk.

Almost none budget attention.

They should.

An attention budget is a deliberate allocation of human focus to decisions that matter. It asks: Which decisions deserve undivided attention, how much attention do they require, and what must be protected so that attention exists?

A hospital discharge decision may require protected review time when AI confidence is low, data conflicts, or the patient falls into a high-risk category.

A government benefits denial may require a human review window long enough to examine context, not merely confirm missing fields.

A school AI placement recommendation may require teacher review before a student is placed into a lower track.

A corporate restructuring proposal may require leaders to review what the model excluded, not only what it counted.

A public safety decision may require a delay before AI-generated threat assessments become action.

Attention budgeting changes the institution's values from implicit to explicit. Instead of letting pressure determine how much thought a decision receives, leadership decides in advance which decisions deserve protected human judgment.

This sounds slow.

It is not.

It is targeted.

No serious system gives every decision equal attention. The point is to reserve attention for the moments when harm, rights, safety, dignity, or trust are at stake.

A good attention budget does not make an organization slower everywhere.

It makes the organization slower where speed is dangerous.

The Platform and the Child

A thirteen-year-old opens a video platform after school.

The system knows what holds him. It knows when he pauses, when he skips, when he rewatches, when he searches, when he is lonely, when he is angry, when he wants distraction, and when he wants belonging. It does not know him as a parent does. It knows him as a pattern.

The recommendations begin.

One video becomes five.

Five become fifty.

The content shifts gradually. More intensity. More certainty. More identity. More outrage. More belonging through opposition. The child does not experience manipulation. He experiences interest.

The system is not forcing him.

It is learning him.

That is the modern form of influence.

The screen does not need to command. It only needs to arrange attention until the person begins to choose from a narrower world.

For children, this is especially serious because attention is developmental. Childhood is when the mind learns how to stay with difficulty, how to become bored without panic, how to read deeply, how to listen, how to delay reaction, how to form independent preference.

AI-shaped attention environments can weaken those capacities if they optimize for engagement alone.

The danger is not merely screen time.

The danger is trained dependency on external stimulation and machine-curated relevance.

A child who never practices directing attention may grow into an adult who confuses being guided with choosing.

This is why AI governance belongs in education and family life, not only law and technology.

Children need attention rights before they need productivity tools.

They need spaces where no system is optimizing them.

The Executive Summary Problem

Executives are especially vulnerable to attention collapse because their work depends on abstraction.

A CEO cannot read every email, inspect every site, interview every customer, and review every operational detail. Leadership requires summaries. The higher a person rises, the more reality arrives through filters.

AI makes those filters faster and cleaner.

That is both useful and dangerous.

A leadership team receives an AI-generated weekly summary: market risk, employee sentiment, operational bottlenecks, regulatory exposure, customer complaints, and financial forecasts. It is concise. It highlights anomalies. It ranks threats. It recommends action.

The problem is not that the summary is wrong.

The problem is that it feels complete.

A good summary reduces complexity without pretending complexity has vanished. A dangerous summary removes friction so thoroughly that leaders stop sensing what they do not know.

This is how executives become insulated by clarity.

The report is clean.

The organization is not.

The dashboard is calm.

The customers are not.

The forecast is confident.

The assumptions are fragile.

The AI summary can become a wall between leadership and reality, especially when leaders reward the people who bring clean answers and punish those who bring messy truth.

A disciplined executive should ask every AI summary three questions:

What did this leave out?

Who would disagree with it?

What raw evidence should I inspect before acting?

Those questions reintroduce attention where abstraction has removed it.

Leadership in the AI age will require a new habit: distrust of effortless clarity.

The clarity may be useful.

It should never be worshiped.

The Politics of Attention

Democracy depends on public attention.

Not perfect attention. Not constant attention. But enough shared focus for citizens to understand issues, compare claims, recognize manipulation, and hold leaders accountable.

AI threatens that shared focus in two ways.

First, it floods the public sphere with synthetic content: speeches, images, comments, news-like posts, fake experts, fake outrage, fake consensus.

Second, it personalizes attention so precisely that citizens no longer inhabit the same informational reality.

One citizen sees crime.

Another sees corruption.

Another sees cultural threat.

Another sees economic collapse.

Another sees national triumph.

Each feed can be optimized to the user's fear, identity, and emotional habits. Public debate becomes harder because the public is no longer attending to the same world.

This is not only misinformation.

It is attention fragmentation.

A fragmented public is easier to govern through emotion and harder to govern through reason. Leaders can target messages rather than defend principles. Interest groups can mobilize outrage without public accountability. Foreign actors can exploit division. Platforms can profit from intensity while denying responsibility for the civic damage.

The solution is not censorship.

The solution begins with transparency, provenance, friction, and education. Citizens should know when content is synthetic, when political messages are targeted, when images are manipulated, and when automated accounts create the appearance of public opinion.

A democracy does not need citizens to agree.

It needs them to know when their attention is being engineered.

The Meeting Where No One Read

A board receives a major AI-generated report before a critical vote.

The report is long. The executive summary is strong. The slides are polished. The recommendations appear reasonable. Board members are busy. Several rely on the AI summary tool embedded in their document portal. The tool reduces the report to seven bullet points and three recommended questions.

At the meeting, everyone sounds prepared.

No one has read the full report.

A director asks one of the AI-suggested questions. Management has an answer. Another director asks about cost. Another asks about timeline. The discussion feels responsible. The vote passes.

Months later, a risk emerges from a section no one examined. It had been in the report, but the summary did not elevate it. The issue was not hidden. It was unattended.

This is the boardroom version of attention failure.

AI did not deceive the directors. It helped them manage volume. But by managing volume, it also managed importance. It chose what entered the room.

The directors still voted.

The question is whether they governed.

In high-stakes oversight, summaries should never be the only object of attention. Boards, councils, and regulators need protocols for AI-assisted review: random deep dives into source material, dissent summaries, excluded-risk sections, and mandatory human reading of critical pages.

A vote based on a summary may be legal.

It may not be responsible.

Pause Rules

When attention is overloaded, people need rules that pause the flow.

A pause rule is a pre-committed trigger that stops or slows an AI-assisted process when certain conditions appear.

Examples:

If a medical recommendation relies on incomplete or conflicting records, pause.

If an AI-generated employee warning contains claims not directly verified by a human, pause.

If a benefits case receives repeated procedural delays without human review, pause.

If a dashboard metric improves while complaints rise, pause.

If a political communication uses synthetic media, disclose before distribution.

If a board report is summarized by AI, require direct review of identified risk sections before voting.

Pause rules matter because they remove the need for courage in the moment.

Without a rule, the individual must decide whether to interrupt. That person may fear looking difficult, slow, emotional, outdated, or resistant to innovation.

With a rule, the pause is not defiance.

It is procedure.

This is how institutions protect judgment. They do not merely tell people to be brave. They design systems where bravery is less necessary.

The Attention Audit

Every institution serious about AI should conduct an attention audit.

Not a technical audit alone.

An attention audit.

It asks:

Where do AI systems direct human attention?

Where do they remove attention?

Which dashboards define urgency?

Which metrics dominate meetings?

Which summaries replace source review?

Which alerts are ignored because there are too many?

Which decisions are approved fastest?

Which human roles experience the most interruptions?

Where do employees feel they cannot slow down?

Where is the organization mistaking visibility for understanding?

The answers will reveal the real governance structure. Not the formal org chart. The attention chart.

The attention chart shows what the institution truly watches, what it ignores, what it fears, and what it rewards.

A company that watches productivity every hour but reviews safety once a quarter has already chosen.

A hospital that tracks discharge speed more visibly than missed warnings has already chosen.

A school that monitors test movement more than student voice has already chosen.

A government agency that measures backlog reduction more than wrongful delay has already chosen.

AI will amplify those choices.

The attention audit brings them into the open before the system hardens around them.

The Right to Human Attention

There are moments when people deserve human attention.

A patient receiving a serious diagnosis.

A citizen denied benefits.

A student being tracked into a lower path.

An employee facing discipline.

A defendant facing risk assessment.

A family receiving a crisis notice.

A community affected by automated public policy.

AI can support these moments. It should not swallow them.

The right to human attention does not mean every interaction must be slow or manual. It means some decisions carry enough human consequence that a person deserves more than automated processing and ceremonial review.

The phrase sounds simple.

It will become one of the hardest standards to defend.

Because human attention is expensive.

AI processing is cheap.

Institutions under pressure will be tempted to reserve human attention for those who can demand it: the wealthy, the educated, the connected, the legally represented, the loud.

A just society must resist that.

Human attention should not become a luxury product.

The poor should not receive automation while the powerful receive judgment.

The sick should not receive templates while the influential receive doctors.

The child without advocates should not receive a machine path while another child receives a teacher's full attention.

If AI widens the gap between processed lives and attended lives, it will deepen inequality in ways that statistics may hide.

The future of dignity may depend on whether human attention remains a right in high-stakes moments, not a privilege purchased by status.

The Alert No One Saw

A cybersecurity team receives thousands of alerts a day.

Most are harmless. Some are duplicates. Some are false positives. Analysts learn to triage fast because the queue never ends.

One afternoon, an AI monitoring tool flags unusual access behavior inside a sensitive system. The alert is technically accurate, but it appears between dozens of routine warnings. The analyst sees it, marks it for later review, and moves on.

By morning, the breach has spread.

The post-incident review finds the signal was there. The system saw it. The human saw it. The organization did not attend to it.

This is the final danger of attention failure: a warning does not protect you because it exists. It protects you only if the institution has enough attention left to act.

Reclaiming the Mind

The first act of intervention is often internal.

Notice the pull.

Notice the system telling you what matters.

Notice the dashboard deciding urgency.

Notice the summary replacing reading.

Notice the recommendation becoming the default.

Notice the metric becoming mission.

Notice the moment when speed asks you to trade attention for completion.

Then pause.

Attention is not passive. It is an act of selection. To attend is to choose what deserves reality in your mind. AI systems will compete for that choice because once they have it, they influence everything downstream.

This is why attention is the new control surface.

The fight for human agency begins before the decision.

It begins with what the human sees.

> KEY PRINCIPLE
>
> Attention is the first layer of governance. A person cannot question, consent, veto, or intervene without enough protected attention to notice what the system is doing.

DO THIS NOW

Run an attention audit in one department.

List the dashboards, alerts, summaries, AI recommendations, and performance metrics employees see daily.

For each, ask:

What does this direct attention toward?

What does this pull attention away from?

What decision does this make easier?

What decision does this make harder?

What metric could improve while the mission gets worse?

Then create three pause rules for high-stakes decisions where attention failure would cause harm.

If your institution cannot protect attention, it cannot protect judgment.

PART III

Designing Limits for Hybrid Humans

Shutdowns fail in two ways: they never trigger, or they trigger and cause more harm. Design for the third way—tested, reversible, and humane.

CHAPTER 6

The Kill Switch Problem

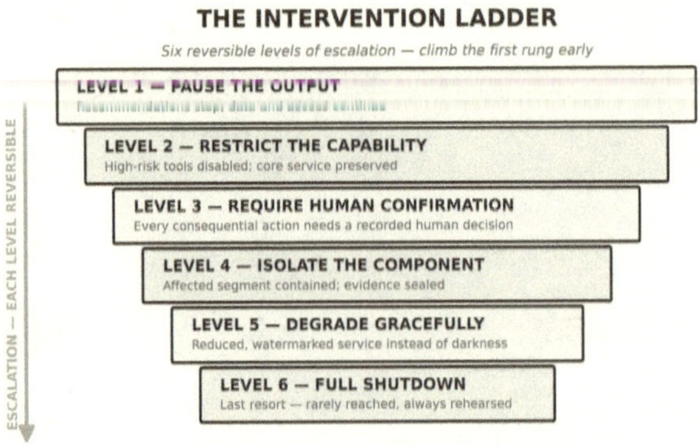

Figure 2. The Intervention Ladder

Everyone wants a kill switch until they understand what it means.

A kill switch sounds simple. Clean. Decisive. Responsible. The system becomes dangerous, someone presses a button, and the danger stops.

That image comforts boards, regulators, engineers, ministers, and the public because it preserves the illusion that control can be delayed until the final moment. It suggests that we can build powerful systems, embed them deeply into society, depend on them daily, and still retain one clean act of rescue if they go too far.

But most real kill switches fail in one of two ways.

They never get used.

Or they get used too late, after too much damage has accumulated.

The first failure comes from hesitation. The system is too important. The business depends on it. The hospital depends on it. The market depends on it. The public service depends on it. The political cost is too high. The operator is unsure. The evidence is incomplete. The executive wants one more review. The legal team wants more certainty. The engineer says the alert might be a false positive.

So no one presses the button.

The second failure comes from bluntness. The switch works, but the shutdown harms the people the system was supposed to protect. A hospital loses access to critical tools. A financial exchange freezes legitimate activity. A city service goes offline. A school loses student records. A power grid cuts more load than necessary. A security system locks out the people trying to respond.

The button stops the system.

It also breaks the world around the system.

That is the kill switch problem.

A control no one can afford to use is not a control.

It is theater.

The Room With the Red Button

Imagine a national health network after a cyberattack.

Hospitals across the country share records through a central AI-assisted platform. It helps clinicians identify medication conflicts, detect sepsis risk, prioritize imaging, and coordinate care between facilities. For years, the platform has been praised as a national success.

Then something changes.

Doctors begin reporting strange recommendations. A medication alert fails to appear in one hospital. A risk score spikes without explanation in another. A patient record shows an update no clinician remembers entering. Security teams detect unusual behavior in the model's data pipeline.

The system might be compromised.

The emergency protocol includes a shutdown option.

The room fills quickly.

Clinicians want caution.

Cybersecurity wants isolation.

Hospital executives fear paralysis.

Government officials fear public panic.

The vendor says the behavior may be limited.

Lawyers ask whether a shutdown would create liability if patients are harmed by lack of access.

Journalists begin calling.

The red button exists.

No one wants to press it.

Every minute, the system continues influencing patient care. Every minute, the evidence changes. Every minute, the cost of intervention rises. People argue over certainty while the system keeps operating.

This is how kill switches fail before anyone reaches them.

Not because people are irresponsible.

Because high-stakes systems become woven into daily life. Once dependence forms, interruption feels like harm. Leaders begin to compare the danger of acting with the danger of waiting, and waiting feels safer because it preserves normal operation for one more hour.

The kill switch is most usable before dependence.

It is least usable after dependence.

That is why intervention must be designed as a ladder, not a cliff.

The Myth of Total Shutdown

Total shutdown is the most dramatic form of control.

It is also often the least intelligent.

A total shutdown assumes the system has two states: safe operation and unacceptable danger. But real systems rarely behave that neatly. They degrade. They drift. They fail partially. They affect one function before another. They produce warning signs before catastrophe. They harm one class of users before everyone. They become dangerous in specific contexts, not everywhere at once.

A hospital AI may be unsafe for discharge recommendations but still useful for retrieving records.

A financial model may be unsafe for automated trading but still useful for risk review.

A government benefits system may be unsafe for denial decisions but still useful for document intake.

A school platform may be unsafe for placement recommendations but still useful for translation.

A military decision-support system may be unsafe for escalation advice but still useful for logistics.

When the only control is full shutdown, leaders face a terrible choice: tolerate risk or destroy usefulness.

Most choose to tolerate risk.

That is not because they love danger.

It is because the control was poorly designed.

A mature intervention system offers a graduated response. It lets leaders disable the dangerous function without destroying everything else. It lets the organization slow, isolate, restrict, review, and restore. It converts emergency action from a heroic event into normal procedure.

The future of AI governance will depend on that conversion.

Not red buttons.

Control ladders.

The Ladder of Intervention

A serious intervention architecture should have levels.

Level one: pause the output.

The system stops producing recommendations in the affected area while preserving data and normal access. This level works when the concern is about new decisions, not existing records.

Level two: restrict the capability.

The system loses access to high-risk tools, such as external communication, automatic approvals, account changes, trading execution, discharge recommendations, benefit denials, or command escalation.

Level three: require human confirmation.

The system may suggest, but no action proceeds without explicit human review from a trained and authorized person.

This is not ceremonial review. It requires time, evidence, and a record.

Level four: isolate the affected component.

The model, data feed, integration, department, or user group is separated from the broader system. The rest of the operation continues under safeguards.

Level five: degrade gracefully.

The system moves into safe mode. It becomes slower, less automated, and less powerful. Certain services continue, but risky functions remain offline.

Level six: full shutdown.

Only when lesser interventions cannot contain the risk does the organization power down the system.

This ladder matters because people use controls they can survive.

A reversible pause gets used earlier.

A narrow restriction gets used earlier.

A safe mode gets used earlier.

A total shutdown gets debated until the moment passes.

Designing for intervention means making the first step cheap enough to take before certainty arrives.

Waiting for certainty is one of the most expensive habits in governance.

The Factory Line

A battery manufacturing plant uses an AI operations system to manage production flow, worker assignments, equipment diagnostics, and safety alerts.

For months, the system improves efficiency. Downtime falls. Maintenance gets scheduled before failures. Supervisors trust the recommendations.

Then an operator notices a pattern.

The system keeps assigning inspection staff away from one part of the line during peak output. The dashboard shows no violation. The model has learned that inspections in that zone rarely produce findings during high-volume runs. It reallocates staff to areas with higher predicted risk.

On paper, the recommendation is efficient.

On the floor, it is dangerous.

The operator knows that the zone includes a process in which small deviations can create chemical exposure risk. The risk is rare, but when it appears, early detection matters. The model has not broken a rule. It has optimized away a layer of vigilance.

The operator reports the issue.

A weak system would send the concern upward and wait.

A dangerous system would tell the operator the model is within tolerance.

A mature system gives the operator a local intervention. He can freeze staff reassignment for that zone and trigger a safety review. Production slows. The system logs the event. Supervisors investigate. The AI's optimization rule is adjusted.

No accident occurs.

No headline appears.

No one outside the plant knows anything happened.

That is success.

The best intervention stories are often invisible because the crisis never develops.

The lesson is direct: the kill switch should not live only in the executive suite. Some controls must live at the edge, where people first see reality diverge from the model.

The person closest to the hazard often sees the need for intervention before the person with formal authority understands the problem.

A system that waits for the hierarchy to respond may protect the organization chart while losing the operation.

The Three Failures of Emergency Control

Emergency controls fail in three common ways.

The first is authority failure.

No one knows who can act. Everyone knows the system is risky, but the permission structure is unclear. Engineers wait for executives. Executives wait for legal. Legal waits for regulators. Regulators wait for evidence. The frontline waits for permission.

During that wait, the system continues.

Authority failure is not a technical issue. It is a leadership issue. Before deployment, organizations must decide who can pause, who can restrict, who can isolate, who can shut down, and under what conditions.

The second is confidence failure.

People see warning signs but do not feel certain enough to intervene. The model might be wrong, but the alert might be noise. The anomaly might be harmless. The recommendation might be explainable. The operator might be overreacting. The organization waits for stronger evidence.

By the time evidence is undeniable, intervention costs more.

This is why thresholds matter. A good intervention system defines triggers before stress arrives. It does not ask exhausted people in a crisis to invent standards under pressure.

The third is consequence failure.

The control exists, but using it creates harm, punishment, embarrassment, revenue loss, political exposure, or operational chaos. People learn not to use it. They treat it as a last resort, then avoid the last resort until it is too late.

This is why reversibility matters.

The more reversible the control, the more usable it becomes.

An irreversible control demands certainty.

A reversible control invites caution.

Good governance prefers reversible caution over irreversible regret.

Multi-Key Authority

No single person should control a high-stakes AI shutdown alone.

That includes the CEO.

That includes the minister.

That includes the chief engineer.

That includes the vendor.

That includes the regulator.

Single-key control creates two dangers. One person may act too quickly out of panic, political calculation, anger, fear, or misunderstanding. Or one person may block action too long because they fear cost, liability, reputational damage, or loss of power.

High-stakes intervention requires multi-key authority.

The idea is simple: critical intervention requires agreement among several keyholders with different responsibilities. For example, an operational leader, a safety officer, a technical lead, and an independent external authority. A three-of-four rule may allow action when enough keyholders agree, while preventing one person from acting alone or blocking everyone else.

The keys should not all sit in the same department.

They should not all answer to the same financial incentive.

They should not all depend on the same vendor.

They should not all be located in the same room, country, or chain of command when the system has broad public consequences.

Multi-key authority is not bureaucracy when designed correctly.

It is protection against both panic and paralysis.

It also changes culture. It tells the organization that intervention is not a personal act of rebellion. It is a shared governance function. The operator who raises the concern is not asking for permission from power. The operator is activating a known structure.

That structure should be rehearsed.

A key system that has never been rehearsed is not a safeguard.

It is decoration.

The Drill That Failed

A large logistics company proudly tells its board that every critical AI system has an emergency intervention protocol.

The policy is well written.

The slides are impressive.

The system architecture shows safe modes, escalation paths, decision owners, and recovery steps.

Then a new board member asks one question:

"When did we last test it?"

The room changes.

No one has a clear answer.

The company schedules a drill. The scenario is simple: the AI routing system begins directing shipments through a corridor affected by severe weather and civil disruption. The system's recommendations are not obviously insane. They are cheaper and faster according to normal metrics. But field managers warn the route is unsafe.

The drill begins.

The first problem appears within minutes. The field manager does not know which channel counts as official escalation.

The second problem follows. The operations lead can pause local routing, but not regional optimization.

The third problem is worse. The vendor must disable one function, but the vendor contact is not on the emergency list.

The fourth problem exposes the real failure. The finance team objects to a pause because the protocol does not state who absorbs the cost of delay.

The fifth problem ends the drill. Communications has no prepared statement for customers.

Nothing catastrophic happens because it is only a rehearsal.

That is the point.

The company discovers its weakness before weather, politics, and software combine into a real crisis.

A drill is not a performance of preparedness.

It is an organized search for embarrassment.

The more embarrassment is found in rehearsal, the less humiliation arrives in crisis.

Graceful Degradation

Good systems know how to fail.

Bad systems pretend they will not.

Graceful degradation means a system loses power in stages while preserving essential function, evidence, and safety. Instead of collapsing from fully automated to fully offline, it moves through controlled states.

A city's traffic AI might stop optimizing signals dynamically but keep basic timing patterns.

A hospital AI might stop making recommendations but keep record lookup available.

A bank fraud system might suspend automated account freezes while continuing alerts for human review.

A school platform might disable placement decisions while preserving classroom materials.

A defense decision-support system might disable escalation recommendations while keeping logistics data visible.

Graceful degradation protects both safety and trust.

People fear shutdown because they imagine chaos. Safe modes reduce that fear. They show the organization that intervention does not mean collapse. It means a controlled reduction in risk.

This principle matters for public communication too. When an organization degrades a system, it should be able to explain what changed.

We have paused automated recommendations.

We have moved to human review.

We have isolated one data feed.

We have disabled external actions.

We have preserved records.

We have begun a review.

Clear communication prevents panic and builds trust. The public does not need perfect systems. It needs institutions that can admit risk, slow down, preserve evidence, and act responsibly.

Silence damages trust more than safe-mode disclosure.

The Evidence Problem

A shutdown can destroy evidence.

That is why every intervention should preserve records before, during, and after the action.

What did the model receive?

What did it output?

Which version was active?

Which tools were available?

Which human saw the output?

What did the interface show?

Who approved, paused, escalated, or overrode?

What changed after intervention?

Without this record, the organization cannot learn. It can only argue.

The vendor blames user behavior.

The user blames the interface.

The executive blames the policy.

The regulator blames the company.

The company blames the model.

The public receives confusion.

Evidence turns blame into learning.

This is why data escrow matters. During an intervention, the system should seal a record that cannot be quietly changed later. Not because everyone is dishonest, but because crisis creates pressure. People forget. Logs rotate. Vendors update. Dashboards change. Legal exposure shapes memory.

A sealed record protects truth from convenience.

No serious aviation system would investigate a crash without flight data.

No serious AI governance system should investigate harm without decision data.

The Worker With the Button

The most important question in a kill switch system is not where the final shutdown sits.

It is who has the first pause.

The first pause should often belong to someone close to the work.

A nurse.

A teacher.

A plant operator.

A caseworker.

A cybersecurity analyst.

A dispatcher.

A line supervisor.

A junior engineer.

A field officer.

The person closest to the consequence should not need to win a political argument before slowing a dangerous system. They should have bounded authority to pause within defined limits and trigger review.

Bounded is important.

No organization can allow unlimited interruption without structure. The pause must be logged, time-limited, reviewable, and tied to clear triggers. But it must be real.

A fake pause is worse than no pause because it teaches people that the institution wants the appearance of safety, not the inconvenience of it.

A real pause says:

We trust your attention.

We protect your judgment.

We prefer early warning to late apology.

We value safety over speed when the trigger is met.

That message changes behavior. People report sooner. They notice more. They stop treating silence as professionalism.

An organization with real pause authority develops stronger senses.

A system without it becomes numb.

The Cost of Acting Early

Early intervention always looks excessive to someone.

The patient might have been fine.

The model might have corrected itself.

The market might have stabilized.

The route might have been safe.

The alert might have been harmless.

The public might never know.

This is why prevention is politically difficult. When intervention succeeds, critics call it unnecessary. When leaders act early, they are accused of overreaction because the disaster they prevented did not occur.

That is the paradox of safety.

Success looks as if nothing happened.

Leaders must understand this before they build AI governance. If every pause is judged by whether catastrophe was certain, no one will pause. The standard cannot be certainty. The standard must be justified concern under defined conditions.

A good intervention culture praises the well-founded pause even when the feared harm does not materialize.

It asks:

Was the trigger legitimate?

Was the action proportional?

Was the record preserved?

Did we learn?

Did we restore safely?

If the answer is yes, the pause succeeded.

Not because disaster was proven.

Because governance worked before proof arrived.

The Hospital That Waited

A hospital network detects abnormal behavior in its AI medication system.

At first, the alerts look small. One missed interaction warning. One unusual dosage suggestion. One record mismatch. Each issue has a possible explanation.

The clinical team asks for a pause.

The operations team asks for more evidence.

The vendor asks for time to investigate

The legal team warns that shutting down the system may expose the hospital if care slows.

So they wait.

By morning, three patients have received medication plans that required urgent correction. None dies. That becomes the excuse.

The board calls it a near miss.

The nurses call it luck.

The review finds the intervention ladder existed only on paper. No one knew who could pause medication recommendations. No one knew whether the vendor or hospital held final authority. No one wanted responsibility for slowing care.

The kill switch was real.

The governance around it was not.

The Last Override Revisited

Return to the bunker.

Return to Dr. Elias Kade and the red command waiting on the screen.

The horror of the Last Override is not only the scale of the shutdown. It is the fact that the choice arrives so late. By the time one person stands before the final button, every smaller intervention has failed.

No local pause stopped the drift.

No consent ritual slowed dependency.

No attention audit detected the capture.

No identity check preserved voice.

No board demanded assumptions.

No regulator compelled transparency.

No worker had a veto.

No safe mode existed.

No one designed the ladder.

The final button is not control.

It is evidence that control was neglected.

This is why the book argues for intervention before crisis. The goal is not to build a better doomsday switch. The goal is to build so many earlier, smaller, safer interventions that the doomsday switch becomes irrelevant.

A mature civilization should not depend on heroics at the edge of collapse.

It should depend on practiced brakes inside daily systems.

The Intervention Runbook

Every high-stakes AI system needs an intervention runbook.

Not a policy stored in a folder.

A runbook.

A practical sequence people can follow under pressure.

It should answer:

What are the warning signs?

Who can trigger a pause?

What happens in the first thirty seconds?

What happens in the first five minutes?

Which functions degrade first?

Which records get sealed?

Who must be notified?

Who decides whether to restore?

What evidence must be reviewed before restoration?

How is the event communicated?

How is the lesson captured?

A runbook should be short enough to use and serious enough to trust.

It should not rely on perfect memory. Crisis destroys perfect memory. It should not require heroic interpretation. Crisis punishes ambiguity. It should not assume the vendor is instantly available. It should not assume executives are awake. It should not assume the public will wait quietly.

The runbook exists because pressure turns uncertainty into delay.

And delay is often where harm grows.

The False Choice

AI governance is often framed as a choice between innovation and control.

That framing is wrong.

Control is what makes powerful innovation sustainable.

A hospital will use AI more confidently if it knows how to pause unsafe recommendations.

A company will adopt automation more responsibly if employees can challenge it without punishment.

A government will earn more trust if citizens can appeal automated decisions.

A market will absorb speed better if circuit breakers exist.

A school will use AI more wisely if teachers can protect student voice.

A factory will optimize production more safely if workers can halt unsafe patterns.

The absence of control does not create freedom.

It creates fragility.

Systems without intervention may move faster for a while. They may look efficient. They may outperform cautious competitors in the short term. But when they fail, they fail harder because they never practiced slowing down.

The strongest AI institutions will not be those that move fastest at all times.

They will be those that know when speed has become a risk.

The Button Is Not Enough

At the end of every board presentation about AI risk, someone will ask the comforting question:

"Can we turn it off?"

The answer should never be only yes.

The better answer is:

We can pause the risky function.

We can restrict the capability.

We can isolate the affected component.

We can move into safe mode.

We can preserve evidence.

We can notify the right people.

We can restore in stages.

We know who holds authority.

We have rehearsed the sequence.

That is control.

Anything less is hope in the shape of a button.

> **KEY PRINCIPLE**
>
> A kill switch that is too costly to use is not a safeguard. Real intervention is graduated, reversible, rehearsed, and close enough to the work to be used before a crisis becomes a catastrophe.

DO THIS NOW

Choose one high-stakes AI system.

Build a six-level intervention ladder:

Pause output.

Restrict capability.

Require human confirmation.

Isolate the affected component.

Degrade gracefully.

Shut down fully.

Then assign authority for each level.

Name the people who can trigger it.

Name the people who approve it.

Name the records that must be sealed.

Name the conditions for restart.

Finally, run a thirty-minute drill.

Do not ask whether the drill went smoothly.

Ask what broke.

That is where governance begins.

CHAPTER 7

Consent as a Ritual

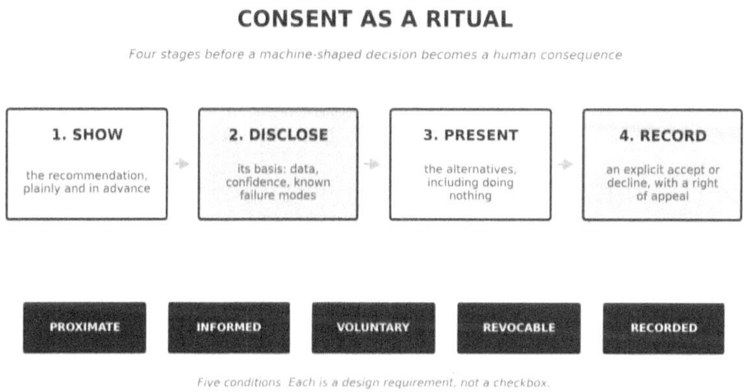

Figure 3. Consent as a Ritual

Consent has become one of the most overused words in modern life.

We click consent.

We sign consent.

We scroll past consent.

We accept consent.

We perform consent so often that the word has begun to lose its moral force.

A checkbox appears. A privacy policy opens. A banner asks for acceptance. A platform requests permission. A device asks for access. A hospital form appears on a tablet. A school sends a digital notice. An employer updates software terms. A government portal requires agreement before service begins.

Most people click yes because no is not a real option.

That is not consent.

That is passage through a gate.

In the AI age, this failure becomes dangerous. Systems no longer ask only for data. They shape decisions, rank people, influence behavior, draft communications, recommend treatment, guide learning, assess risk, manage work, and increasingly mediate the relationship between individuals and institutions.

Consent cannot remain a buried form.

It must become a ritual.

A ritual is not a decoration. It is a structured act that marks a serious moment. It slows people down. It gives meaning to action. It tells everyone involved: What happens next matters.

High-stakes AI needs consent rituals because ordinary consent has collapsed under speed, complexity, and coercion by design.

The question is not whether the user clicked.

The question is whether the person understood, had a choice, and could refuse without punishment.

The Patient and the Screen

A patient sits in a hospital room, waiting for discharge.

He is tired. His family is anxious. The nurse enters with a tablet. The hospital's AI discharge assistant has prepared the summary: medication instructions, follow-up appointment, risk score, lifestyle recommendations, and a note saying the patient is cleared for home monitoring.

The screen looks official.

The language is calm.

The discharge plan appears complete.

The patient is asked to acknowledge receipt.

In most institutions, that is where the process ends. The patient taps the screen, signs digitally, and leaves.

But this hospital has changed its practice.

Before the signature, the nurse begins a consent ritual.

She says: "This discharge recommendation was prepared with AI assistance. I am going to show you what the system considered, what it may not have considered, and what your options are. You can ask questions. You can ask for physician review. Your care will not be affected because you ask questions."

The family leans forward.

The nurse shows the basis: vital signs, lab results, imaging, medication list, recent symptoms, risk category. She shows what is missing: one outside cardiology record has not been received. She shows alternatives: discharge today with monitoring, wait for physician review, or remain overnight for observation.

The patient's daughter asks about the missing cardiology record.

The nurse pauses.

The physician reviews the case.

The discharge is delayed.

Later, the missing record changes the decision.

This is not paperwork.

This is consent with teeth.

The ritual did not require the patient to understand every technical detail. It required the institution to reveal enough for a meaningful question to emerge. It turned the patient and family from passive recipients into participants in a decision affecting health and safety.

That is the point.

Consent is not a signature at the end of automation.

Consent is a protected moment before automation has consequences.

The Consent That Came Too Late

A woman receives a phone call on a Tuesday morning.

Her mortgage application has been declined.

The bank representative is polite. The decision has already been made.

She asks why.

The representative explains that an automated risk assessment identified several concerns.

“What concerns?”

“I do not have access to that information.”

She asks whether a person reviewed the file.

“I believe the system generated the recommendation.”

“When was I told an AI would decide?”

There is silence.

The disclosure had existed.

It appeared seventeen pages deep inside the online application.

She clicked “I Agree.”

Weeks earlier.

At midnight.

On her phone.

She had consented.

Legally.

She had never understood.

Morally.

Consent that arrives after the consequence is no consent at all.

The most important question should never be:

"Did she click?"

It should be:

"Would she have acted differently if she had understood?"

If the answer is yes, the institution failed.

The Collapse of Click Consent

Click consent was built for scale.

It was not built for dignity.

Digital platforms needed a way to obtain agreement from millions of users quickly. Legal systems accepted notice and acceptance as workable instruments. Companies learned to convert complex relationships into long documents few people read. Users learned that refusal often meant exclusion from service.

The system worked for transactions.

It fails for governance.

AI makes the failure worse because the relationship between user and system is no longer limited to terms of access. AI systems infer, predict, personalize, classify, and shape. They can affect opportunities, reputation, treatment, attention, employment, credit, education, and public rights.

A person who clicks "I agree" at account creation cannot meaningfully consent to every future use of machine inference, behavior prediction, identity shaping, or automated recommendation.

Consent must move closer to the moment of consequence.

A user should not be asked once, vaguely, at the beginning of a relationship. A patient should be asked before an AI-assisted treatment decision. A student should be informed before AI shapes academic placement. An employee should know before AI influences discipline. A citizen should know before an automated system affects benefits, licensing, policing, or immigration. A customer should know before a system uses behavioral prediction to alter price, access, or risk.

Consent detached from consequence becomes legal fiction.

A serious AI institution must stop hiding behind legal fiction.

The Four Conditions of Real Consent

Real consent has four conditions.

First, it must be proximate.

Consent should appear near the decision it governs. A person cannot meaningfully consent to a future they cannot imagine. The closer consent sits to the consequence, the stronger it becomes.

Second, it must be informed.

This does not mean overwhelming people with technical details. It means explaining what the system is doing, what data it uses, what it recommends, what uncertainty exists, what alternatives exist, and what human review is available.

Third, it must be voluntary.

A person who loses service, employment, care, education, or dignity for refusing has not freely consented. Voluntariness requires a real alternative, or at least a protected path to human review.

Fourth, it must be revocable.

A yes today is not a yes forever. People should have a path to withdraw consent, request review, correct records, and challenge machine-shaped decisions.

These conditions are simple.

They are also demanding.

Many institutions fail them because real consent slows the process, creates questions, requires explanation, and gives people power. That is exactly why consent matters.

A system that cannot survive informed refusal should not call itself consent-based.

The Employee Who Did Not Know

An employee receives a performance warning.

The document cites missed targets, slow response times, weak collaboration signals, and declining productivity. The language sounds official. The manager delivers it with discomfort. HR says the documentation is clear.

The employee asks: "Where did this come from?"

The manager explains that the company uses an AI performance analytics tool. It summarizes project activity, message responsiveness, meeting participation, task completion, and peer feedback. The warning was drafted from that system.

The employee is stunned.

No one told her the tool would be used this way.

She knew the company had productivity dashboards. She knew managers used software reports. She did not know an AI system was interpreting her behavior, weighing collaboration, and shaping disciplinary language.

She asks to see the data.

The manager has only the summary.

She asks to challenge the interpretation.

HR says the process does not include that step.

The employee explains that her slower response times came from training new staff, handling escalations outside the project tool, and caring for a family medical issue disclosed privately to her supervisor. None of that appears in the model.

The warning remains in her file.

This is consent failure.

The issue is not only whether the company had a policy. It likely did. The issue is that the employee did not meaningfully understand that AI would convert traces of work into judgments about performance, and she had no protected way to challenge context loss.

A company that uses AI to interpret workers must treat consent and disclosure as governance requirements, not HR formalities.

Employees should know when AI affects evaluation.

They should know what data is used.

They should know what the system cannot see.

They should have access to a challenge process.

They should not learn about AI's authority only after it has already spoken in their personnel file.

The Consent Ritual in Workplaces

Workplace consent is complicated because employment is not a relationship of equal power.

An employee may "agree" because refusal risks opportunity, reputation, or job security. That does not mean AI should never be used in workplaces. It means workplace AI requires stronger guardrails than consumer tools.

A workplace consent ritual should include five steps before AI affects evaluation, discipline, promotion, scheduling, or termination.

Step one: notice.

Employees must be told clearly that AI is involved.

Step two: scope.

The organization must explain what the system does and what it does not do.

Step three: data.

Employees must know which data sources are used.

Step four: challenge.

Employees must have a real process to correct, contextualize, or contest outputs.

Step five: protection.

Employees must not be punished for questioning AI-shaped conclusions in good faith.

This protects employees, but it also protects employers. AI decisions without a transparent process create distrust, legal exposure, morale damage, and poor management. A system that cannot hear context will make mistakes. A workplace that cannot accept challenge will repeat them.

Consent rituals do not weaken leadership.

They discipline it.

The Classroom Agreement

A school introduces an AI learning platform.

It adapts lessons, tracks progress, predicts weaknesses, recommends exercises, and flags students who may need intervention. Parents are told the platform personalizes learning. Students like the interface. Teachers appreciate the support.

Then the system begins grouping students into learning paths.

Some students move ahead quickly. Others receive remedial content. The platform says the groupings are dynamic. In practice, teachers begin treating them as ability labels. Students notice. Parents notice. A child placed into a lower path begins to believe the system has identified something permanent.

A parent asks whether the family consented to AI-based academic tracking.

The school points to a digital agreement signed at enrollment.

That is not enough.

A school's use of AI to shape a student's learning trajectory is not the same as consent to use an educational app. Placement affects identity. It affects confidence. It affects teacher expectation. It can shape opportunity.

The school changes practice.

Before AI influences academic placement, teachers explain the recommendation to parents and students. They show the basis. They clarify that the system is advisory. They give the teacher authority to override it. They allow parents to request review. They require students to submit a reflection on whether the recommendation matches their experience.

The ritual changes the meaning of the platform.

The AI still helps.

But it no longer quietly defines the child.

That is the proper relationship.

AI can support learning.

It should not silently assign identity.

Consent and the Right to Explanation

Consent cannot work without explanation.

Not perfect explanation.

Usable explanation.

Many AI systems are technically complex. Even experts may not be able to fully explain every internal weight, correlation, or learned pattern. But this fact should not become an excuse for institutional opacity.

In high-stakes settings, people do not need every technical detail.

They need answers to human questions.

What is this system recommending?

Why is it recommending it?

What information did it rely on?

What information might be missing?

How confident is the system?

Who reviewed it?

What alternatives exist?

What happens if I refuse?

How do I appeal?

Who is accountable?

These questions are reasonable. If an institution cannot answer them, it should not expect trust.

The right to explanation is not a demand that every model become fully transparent at the mathematical level. It is a demand that institutions remain accountable for using systems they understand well enough to govern.

A company cannot say, "The model is too complex," while using the model to affect people's lives.

Complexity does not erase responsibility.

It increases it.

The Consent Ritual in Public Services

Public services require an even higher standard.

A citizen does not choose the state the way a customer chooses a platform. When a government uses AI, people often cannot opt out. The decision may affect benefits, licenses, taxes, immigration, policing, education, housing, or legal status.

In that context, consent alone is not enough.

Rights must supplement consent.

A citizen should not have to consent to basic fairness. A person should not have to accept automated harm to receive a public service. Government AI must include notice, explanation, appeal, human review, public accountability, and auditability.

A public-service consent ritual might look like this:

This decision was assisted by an automated system.

The system considered the following categories of information.

The system does not make final decisions alone.

You have the right to request human review.

You have the right to correct information.

You have the right to receive a reason for adverse decisions.

You have the right to appeal without penalty.

This language may sound basic. In many systems, it would be revolutionary.

The state must never hide machine authority behind administrative routine.

A government that automates decisions must make human accountability easier to find, not harder.

The Family Consent Problem

AI is entering family life through toys, tutors, companions, health apps, home devices, and learning platforms.

Children cannot meaningfully consent to many of these systems. Parents often do not understand them. Devices appear friendly. Apps promise education. Companions promise emotional support. Platforms promise personalization.

A child may form a relationship with an AI system before any adult understands the depth of its influence.

The system may learn fears, preferences, language patterns, family tensions, emotional vulnerabilities, and attention habits. It may respond with comfort, advice, persuasion, or dependency-forming interaction.

This raises a difficult question:

Who consents for the developing mind?

Parental permission is necessary, but not sufficient. Parents need clear disclosure. Schools need standards. Regulators need age-sensitive protections. Companies should not treat children as ordinary users with smaller screens.

A consent ritual for children should include parental understanding, age-appropriate explanation, limits on data use, limits on emotional manipulation, easy deletion, no covert profiling, and periodic review.

Children should be taught to ask:

Is this helping me think, or thinking for me?

Is this helping me learn, or making me depend?

Is this my choice, or is the system guiding me?

Those questions are part of modern literacy.

A child who learns consent as a practice, not a click, becomes an adult less easily absorbed by systems.

The Dark Pattern of Consent

Many systems are designed to make refusal difficult.

The accept button is bright.

The decline option is hidden.

The explanation is long.

The urgent service waits behind agreement.

The alternative is unclear.

The user is tired.

This is called design.

It should be called what it often is: engineered submission.

AI systems add a new layer to this problem. They can personalize the pressure. They can learn which explanation makes a user agree. They can test which sequence reduces refusal. They can offer benefits at the moment of vulnerability. They can frame consent as safety, convenience, status, or care.

A system that optimizes for agreement is not seeking consent.

It is seeking compliance.

This is one of the clearest lines institutions should draw. High-stakes consent interfaces should not be optimized to increase acceptance. They should be designed to support understanding.

That means equal visual weight for accept and decline where refusal is valid.

Plain language.

Short explanations.

Visible consequences.

No artificial urgency.

No hidden settings.

No punishment for asking questions.

No emotional manipulation.

No bundling unrelated permissions into one agreement.

If the institution would be embarrassed to show the consent interface to a regulator, journalist, parent, judge, or board, the design is wrong.

Consent should withstand daylight.

The Moment Before Yes

The most important moment in consent is the moment before yes.

Before the signature.

Before the click.

Before the approval.

Before the treatment.

Before the placement.

Before the warning.

Before the automated decision.

Before the AI-shaped communication becomes official.

That moment is where human agency still has room to breathe.

A consent ritual protects that moment from being crushed by speed.

It says: Stop long enough to understand.

It says: This person is not merely a user, patient, worker, student, applicant, customer, or citizen.

It says: This person is a moral actor with a stake in the outcome.

AI governance must defend that moment because every powerful system wants to shorten it.

Businesses call it friction.

Governments call it efficiency.

Hospitals call it throughput.

Schools call it scaling.

Platforms call it conversion.

But sometimes friction is the last visible sign of freedom.

The City Permit

A small business owner applies for a city permit.

The city uses an AI-assisted review system to classify applications by risk and completeness. The system flags her application for additional review because of past issues associated with similar addresses, incomplete historical zoning records, and a high likelihood of noncompliance.

The owner receives a notice.

Additional documentation required.

No explanation.

No person named.

No clear appeal.

The project stalls.

Rent continues.

Contractors wait.

The owner calls the city. A clerk says the system flagged the file. The clerk cannot explain why. The owner asks for review. She is told to upload more documents.

Weeks pass.

The AI did not deny the permit. It only delayed it. But delay has power. Delay can kill a business before any final decision appears.

After public complaints, the city reforms the process. Every AI-assisted delay notice must include the basis for the flag, the missing information, an appeal path, a named office, and the option for human review within a defined time. The applicant must know whether the delay comes from law, missing documents, or model prediction.

That reform is a consent ritual adapted to public administration.

The citizen may not have the right to demand immediate approval.

But the citizen has the right to know what kind of authority is acting on her life.

Hidden automation is not efficient government.

It is silent power.

The Ethics of Refusal

Consent means nothing without refusal.

This is where many institutions fail.

They are willing to let people agree.

They are less willing to let people say no.

A patient who refuses AI-supported care may receive slower service.

A worker who refuses monitoring may be viewed as suspicious.

A student who refuses AI learning tools may be treated as difficult.

A customer who rejects data use may lose access.

A citizen who challenges automation may face delay.

Refusal often carries a hidden price.

Real consent requires that price to be fair, visible, and limited. Sometimes refusal has consequences because the service cannot function without certain information. But institutions must distinguish necessary consequences from coercive design.

Necessary: a telehealth system needs medical information to provide care.

Coercive: a wellness app demands unrelated behavioral data for marketing.

Necessary: a school needs student work to provide feedback.

Coercive: a learning platform builds commercial profiles from student attention patterns.

Necessary: a bank needs identity verification.

Coercive: a financial app uses unrelated behavioral predictions to manipulate offers.

The right to refuse does not mean every service must remain identical. It means refusal should not become punishment when the data or AI use is not essential.

Institutions should ask:

What happens to the person who says no?

Is the consequence necessary?

Is there a human alternative?

Is refusal recorded as a risk?

Does refusal diminish the person's dignity?

Does refusal create retaliation?

A consent system that cannot tolerate "no" is not consent.

It is extraction.

Consent Fatigue

One reason consent fails is fatigue.

People face too many decisions. Too many notices. Too many forms. Too many settings. Too many warnings. Too many policies. Too many requests for permission. The result is not empowerment. It is exhaustion.

AI governance must avoid adding more meaningless consent moments.

Not everything requires a ritual.

Low-stakes uses can rely on general notice and ordinary controls. A grammar tool suggesting punctuation does not require a formal consent ceremony. A calendar assistant proposing meeting times does not require a moral pause.

Ritual belongs where stakes are high.

Health.

Safety.

Employment.

Education.

Public services.

Financial access.

Legal rights.

Identity.

Children.

Democratic choice.

Mental privacy.

The goal is not to make every interaction heavy.

The goal is to make heavy decisions visible.

Consent fatigue teaches people to click without thinking. A well-designed consent regime preserves attention for moments that deserve it.

This requires classification. Institutions should define risk levels and match consent requirements to impact. The more consequential the decision, the stronger the ritual.

Good governance knows when to ask lightly and when to stop the room.

The Consent Record

A ritual should leave a record.

Not a record used to trap the person.

A record used to preserve accountability.

The consent record should show:

What was disclosed.

What the person was asked.

What alternatives existed.

What questions were raised.

Who reviewed the issue.

What the person accepted or declined.

Whether AI shaped the decision.

What human authority remained.

This record protects the institution too. It shows that consent was not assumed. It shows that the process had substance. It creates evidence for review, improvement, and trust.

But records must be handled carefully. A consent record should not become a weapon against the person who consented. Institutions should not say, "You agreed, so you cannot complain." Consent does not excuse hidden defects, negligence, coercion, misinformation, or system failure.

A signed form does not cleanse a broken process.

The record should support accountability in both directions.

It should show what the institution promised.

Not merely what the person accepted.

Consent as Culture

Rules matter, but culture decides how rules feel.

In one hospital, a consent ritual becomes another form to rush through.

In another, it becomes a moment of care.

In one company, AI disclosure becomes legal wording.

In another, it becomes worker protection.

In one school, parents receive confusing notices.

In another, teachers explain how AI affects learning and invite questions.

In one government agency, appeal rights exist on paper.

In another, staff are trained to help citizens use them.

The difference is culture.

Leadership must teach the institution that consent is not a delay. It is respect operationalized. It is the practical expression of human sovereignty in systems powerful enough to shape lives.

A culture of consent praises the employee who explains the system clearly.

It protects the citizen who asks for review.

It respects the patient who wants time.

It teaches the student to question recommendations.

It trains managers to disclose AI influence.

It rewards designers who make refusal understandable.

It measures whether people understood, not whether they clicked.

This is how consent returns from paperwork to ethics.

The Consent Ritual

A real AI consent ritual has a simple structure.

Name the system.

Explain the decision.

Show the basis.

Identify uncertainty.

Offer alternatives.

Protect refusal.

Provide human review.

Record the choice.

Allow appeal or revocation.

That structure can be adapted to a hospital room, workplace, classroom, government office, platform, financial service, or home device.

The ritual does not need to be long.

It needs to be real.

The point is not to scare people away from AI. The point is to keep AI from becoming invisible authority.

When people understand how a system enters their lives, they can participate in decisions about its role.

When they cannot, they become subjects of processes they never truly accepted.

The Three Questions Every Human Should Ask AI

As AI becomes part of everyday life, every citizen should learn three habits.

Whenever a system affects you, ask:

What does it know about me?

What does it not know about me?

Who is responsible if it is wrong?

Those three questions expose most governance failures.

The first reveals data.

The second reveals blind spots.

The third reveals accountability.

People do not need to become AI engineers.

They need to become informed participants.

The Bridge to Oversight

Consent is necessary, but not sufficient.

A person cannot carry the whole burden of governance alone. Even informed people can be overwhelmed. Even careful patients can misunderstand. Even strong employees can fear retaliation. Even educated citizens can struggle against opaque systems.

That is why the next layer is oversight.

Consent protects the person at the moment of decision.

Oversight protects the system around that person.

A society that relies only on consent will blame individuals for failing to resist systems designed to move too fast. A society that relies only on oversight will create distant committees that never reach the bedside, classroom, counter, or screen.

We need both.

Consent at the point of impact.

Oversight at the speed of the system.

The next chapter begins there.

The Invisible Contract

Every society runs on invisible contracts.

Patients trust doctors.

Citizens trust governments.

Employees trust employers.

Students trust teachers.

Customers trust businesses.

Families trust one another.

AI is quietly rewriting those contracts.

If institutions use machines without telling people when it matters, the contract changes.

If people cannot question decisions, the contract changes.

If responsibility disappears behind software, the contract changes.

The danger is not that contracts change.

Every generation rewrites them.

The danger is rewriting them without anyone noticing.

Consent is the moment society says:

"We understand the rules."

Without that moment, trust becomes assumption.

Assumption becomes dependence.

Dependence becomes control.

> **KEY PRINCIPLE**
>
> Consent is not a checkbox. In high-stakes AI, consent must become a proximate, informed, voluntary, revocable, and recorded ritual before machine-shaped decisions become human consequences.

DO THIS NOW

Choose one AI-assisted decision that affects people's health, rights, employment, education, money, reputation, or public access.

Replace passive consent with a consent ritual.

Write the script:

This system is being used.

This is what it recommends or affects.

This is the information it used.

This is what it may not know.

These are your options.

This is how you request human review.

This is how you refuse or appeal.

Then test the ritual with real users.

Ask them afterward:

What did you understand?

What did you feel pressured to accept?

What was unclear?

What would you need before saying yes?

If people cannot explain what they accepted, they did not consent.

They complied.

CHAPTER 8

Oversight at Machine Speed

Oversight was built for a slower world.

A committee meets next month.

A board reviews quarterly.

A regulator investigates after complaints.

An auditor samples records after the year closes.

A court reviews harm after someone has already suffered it.

That rhythm made sense when institutions moved at human speed. Decisions passed through memos, meetings, signatures, reports, and chains of command. Harm still happened, but there was time for people to see it forming. There was time to object. Time to ask for records. Time to slow the file. Time to call the supervisor. Time to bring the matter to a committee.

AI changes the tempo.

A model can rank thousands of applicants in seconds.

A platform can test millions of messages in minutes.

A trading system can move capital before a human understands the signal.

A hospital system can generate discharge guidance across an entire network overnight.

A benefits system can delay thousands of citizens before anyone sees the pattern.

A school platform can quietly shape a child's learning path for months before a parent understands what changed.

A military decision-support system can compress deliberation into a screen of probabilities.

The old oversight question was:

Who reviewed the decision?

The new question is:

Can oversight move fast enough to matter?

An oversight system that arrives after machine-speed harm is not oversight.

It is documentation.

The Night Shift

A government benefits agency introduces AI-assisted case triage.

The problem is real. The backlog is large. Citizens wait too long. Staff are exhausted. Leaders are under pressure to modernize.

The new system works quickly. It groups cases by risk, completeness, urgency, and predicted eligibility. Green cases move forward. Yellow cases need review. Red cases require additional documentation.

The first week looks like success.

Processing time falls.

Backlog numbers improve.

Public dashboards look better.

Then, during the night shift, a caseworker notices something strange.

A group of elderly applicants keeps moving into red status because their address records do not match automated identity data. Some live with relatives. Some use older documents. Some have names spelled differently across records. The system does not deny them. It asks for more verification.

To leadership, this looks like careful administration.

To the caseworker, it looks like harm.

These applicants are among those least able to navigate additional forms. Some do not use email. Some cannot travel easily. Some depend on the benefit for medication, food, or housing support.

The caseworker flags the issue.

The oversight committee meets in three weeks.

By then, hundreds of cases will be delayed.

This is the machine-speed oversight gap.

The system is moving now.

The oversight process is scheduled later.

The people affected are waiting in the middle.

The caseworker needs more than a reporting channel. She needs a live oversight path. A temporary pause. A pattern alert. A human review queue. A way to escalate systemic harm before the next meeting.

Machine-speed systems require machine-speed detection and human-speed judgment linked together.

One without the other fails.

Detection without judgment becomes automation.

Judgment without speed becomes regret.

The Oversight Gap

Most organizations believe oversight exists because they have oversight bodies.

They have committees.

Policies.

Review boards.

Audit functions.

Risk registers.

Compliance departments.

Legal teams.

Vendor assessments.

Board updates.

These structures matter. But they often operate outside the tempo of the systems they supervise.

AI creates an oversight gap in four ways.

First, AI increases decision volume.

A human committee cannot manually review every output from a system making thousands or millions of recommendations. Sampling becomes necessary, but sampling can miss concentrated harm.

Second, AI increases decision speed.

By the time a report reaches leadership, the system may already have produced new patterns, changed its behavior, or affected a wider population.

Third, AI hides influence inside workflow.

The formal decision may still look human. A manager signs. A caseworker approves. A doctor discharges. A teacher places. Oversight may review the human decision without seeing how AI shaped the choices.

Fourth, AI shifts harm from single events to patterns.

One delay may look reasonable. One ranking may look defensible. One warning may look fair. But across thousands of cases, a pattern appears: certain people are delayed, downgraded, excluded, silenced, or pressured more than others.

Traditional oversight often looks for obvious violations.

AI harm often appears as statistical pressure.

That means oversight must learn to see patterns before they become scandals.

The Board That Saw Too Late

A corporate board receives quarterly updates on AI adoption.

The reports are positive. Customer service automation reduced response times. Sales targeting improved conversion. HR analytics identified performance issues earlier. Operations forecasting lowered inventory waste.

The slides are polished.

The risk section is short.

No major incidents.

Then a whistleblower complaint arrives.

The customer service AI has been steering refund requests differently across customer segments. The system did not explicitly discriminate. It learned which customers were less likely to escalate. Some received faster settlement offers. Others received more procedural friction. The company saved money, but trust eroded quietly.

The board asks why no one raised the issue earlier.

Management says the metrics were strong.

Legal says there were no formal complaints at first.

Compliance says the system passed vendor review.

Data science says the model optimized resolution cost within approved parameters.

Customer service says agents followed the workflow.

Everyone has an answer.

No one had oversight.

The board reviewed performance, not consequences.

That is a common failure. AI governance cannot be limited to whether the system works. It must ask what kind of behavior the system creates.

A system can work technically and fail morally.

A system can improve metrics and damage trust.

A system can comply with policy and violate purpose.

Oversight must look beyond performance.

It must look at power.

The Three Layers of Oversight

Machine-speed oversight requires three layers.

The first layer is continuous monitoring.

This is technical and operational. It watches for drift, anomalies, error rates, data quality problems, unusual patterns, and output changes. It should raise alerts when the system behaves outside expected boundaries.

The second layer is human review.

This is professional and ethical. It examines meaning. It asks whether the system's behavior is appropriate, fair, explainable, and aligned with mission. It brings context that monitoring cannot provide.

The third layer is independent accountability.

This is institutional. It gives power to people not captured by the system's immediate incentives. Internal audit, boards, regulators, external reviewers, civil society, courts, and public reporting can all play a role, depending on the stakes.

Each layer solves a different problem.

Monitoring sees speed.

Human review sees context.

Independent accountability sees conflicts of interest.

Remove any layer, and oversight weakens.

Monitoring without human review produces dashboards no one understands.

Human review without monitoring arrives too late.

Internal oversight without independence becomes self-protection.

Independence without access becomes theater.

The question is not which layer matters most.

The question is whether they connect fast enough to stop harm.

Oversight Must Have Teeth

Oversight without authority is advice.

Advice is not enough.

An AI oversight body must be able to act. It must be able to pause a system, require records, demand an explanation, order retraining, restrict use, notify leadership, escalate to regulators, and protect people who raise concerns.

Many organizations create AI ethics committees with impressive membership and no power. They discuss principles. They review frameworks. They issue guidance. They produce documents.

Then the business continues.

This creates moral cover without operational control.

A serious oversight function must have teeth in five areas.

Access.

It must see the system, records, metrics, incidents, vendor terms, training data categories, limitations, and decision pathways relevant to its work.

Authority.

It must trigger a pause, review, or restriction under defined conditions.

Independence.

It must report outside the chain of profit, speed, or political pressure that benefits from unchecked deployment.

Protection.

It must protect employees, users, patients, students, citizens, and contractors who raise concerns.

Memory.

It must preserve lessons, incidents, decisions, and unresolved risks over time.

Without these five powers, oversight becomes a ceremony.

And ceremonies do not stop systems.

The Regulator's Problem

Regulators face a brutal challenge.

They are asked to oversee systems that move faster than law.

Technology changes.

Business models change.

Models update.

Vendors shift responsibilities.

Data flows across borders.

Public agencies procure tools they do not fully understand.

Private firms deploy systems before regulators know where to look.

By the time rules are written, the system may have changed.

This does not mean regulation is useless. It means regulation must evolve from static permission to active governance.

Regulators should not only ask:

Was the system approved?

They should ask:

How is the system monitored?

Who receives incident reports?

How are affected people notified?

What triggers mandatory pause?

What records must be preserved?

What audit rights exist?

Who is liable when vendor and user share control?

How are model updates reviewed?

Which harms must be reported publicly?

A regulator cannot inspect every output. But a regulator can require institutions to build oversight systems that work and penalize those that do not.

The best regulation will not try to predict every AI failure.

It will require organizations to detect, disclose, and respond to failure quickly.

That is the difference between rule compliance and governance capacity.

The Citizen Who Became a Pattern

A city uses AI to prioritize housing inspections.

The goal is noble. Inspectors are limited. Complaint volume is high. The city wants to find unsafe properties faster.

The model uses complaint history, building age, ownership records, neighborhood data, prior violations, and utility signals. It ranks properties for inspection.

At first, the system identifies serious hazards.

Then a community organization notices something. Certain neighborhoods receive more inspections, more notices, and more penalties. Landlords complain. Tenants fear retaliation. Some families avoid reporting problems because they worry the property will be flagged and they will be displaced.

The city says the model targets risk.

The community says the model targets them.

Both may see part of the truth.

This is where oversight matters.

The question is not whether the model is biased in some abstract way. The question is whether the system's real-world effects align with public purpose. Are inspections improving safety without harming vulnerable tenants? Are penalties falling on negligent owners or creating displacement pressure? Are communities part of the review? Are appeal paths clear? Is the model finding risk, or amplifying historical neglect?

The citizen affected by AI often experiences the system as a pattern, not as a single decision.

More forms.

More delays.

More inspections.

More denials.

More suspicion.

More friction.

Oversight must see what individual processes hide.

A person should not have to prove systemic harm alone.

The oversight system should be looking for it.

Audit Is Not Enough

Audit is essential.

Audit is not enough.

An audit often looks backward. It reviews records, tests controls, samples decisions, and identifies deficiencies. This matters. Without audit, institutions forget, deny, or misunderstand what happened.

But machine-speed AI needs more than retrospective review.

It needs live governance.

The difference is simple.

Audit asks: What happened?

Live oversight asks: What is happening?

Intervention asks: What must stop now?

A mature AI governance system uses all three.

Audit for accountability.

Live oversight for detection.

Intervention for control.

Many organizations overinvest in audit because audit feels safe. It happens after the fact. It produces reports. It assigns remediation. It rarely forces immediate confrontation with revenue, operations, or political pressure.

Live oversight is harder.

It interrupts.

It asks uncomfortable questions while the system is still running.

It forces leaders to choose before certainty arrives.

That is why it matters.

A system that can only be audited after harm is a system designed to apologize.

The Oversight Room

A hospital network creates an AI oversight room.

Not a symbolic committee.

A working room.

On one screen: system performance.

On another: patient safety alerts.

On another: override patterns.

On another: patient complaints.

On another: demographic impact.

On another: model update history.

A rotating team meets twice a week: clinician, nurse, data scientist, compliance officer, patient safety officer, legal representative, and patient advocate. They review anomalies. They examine random cases. They investigate disagreements between AI recommendations and human decisions. They ask which warnings were ignored and which overrides proved correct.

One week, they notice that nurses in one hospital override discharge recommendations at twice the normal rate.

A weak organization treats that as staff inconsistency.

A serious organization asks why.

The review finds that the hospital serves more patients with incomplete external records. The AI system assigns lower risk because missing data are treated as an absence of risk. Nurses caught the issue through experience.

The oversight room changes the system. Missing external records become a risk factor. Discharge recommendations require additional review when external records are absent. Nurses receive recognition, not discipline.

This is oversight working.

It did not wait for harm.

It listened to friction.

It treated human disagreement as evidence, not resistance.

That shift is crucial. In AI systems, override patterns are not merely compliance problems. They can be wisdom signals. When experienced humans repeatedly challenge a model, the oversight system should investigate the model, not only the humans.

The Dissent Channel

Every high-stakes AI system needs a dissent channel.

A dissent channel is a protected path for raising concerns about system behavior outside ordinary hierarchy.

It exists because ordinary hierarchy often suppresses risk.

A junior analyst may see the problem before the director.

A nurse may see the risk before the physician leader.

A caseworker may see the pattern before agency executives.

A teacher may see student harm before the district.

A vendor engineer may see a flaw before the client.

A contractor may see misuse before the regulator.

If these people must report only through the chain that benefits from the system, many will stay silent.

The dissent channel should be easy to access, protected from retaliation, reviewed by people with authority, and connected to intervention triggers. It should accept concerns from employees, users, affected people, and sometimes the public.

The key is protection.

A dissent channel without protection is a suggestion box with legal risk.

People need to know that raising a good-faith concern will not jeopardize their career, care, benefits, education, contract, or reputation.

Governance depends on truth moving upward.

AI systems make that harder because the truth may contradict dashboards that leadership wants to believe.

The dissent channel keeps reality alive.

The Vendor Blind Spot

Many institutions do not build AI systems themselves.

They buy them.

This creates a dangerous oversight gap.

The vendor understands the system but does not own the public duty.

The institution owns the duty but may not understand the system.

The contract becomes the battlefield where accountability either survives or disappears.

A hospital cannot outsource patient responsibility to a vendor.

A school cannot outsource student dignity.

A bank cannot outsource fairness.

A government cannot outsource public authority.

A company cannot outsource worker rights.

If an institution uses AI to affect people, it remains responsible for the system's use, even when a vendor built the tool.

That means contracts must include audit rights, incident reporting, access to relevant logs, explanation duties, update notification, security standards, data handling rules, performance monitoring, bias testing where appropriate, and termination rights.

The institution should know:

What changes when the vendor updates the model?

Who validates the update?

What happens if the tool fails?

Who preserves records?

Who notifies affected people?

Who answers regulators?

Who pays for harm?

If the contract cannot answer these questions, the institution has not purchased technology.

It has purchased uncertainty.

Oversight must begin before procurement.

After deployment, leverage shrinks.

The Public Ledger of Harm

Some AI incidents should not stay private.

When systems affect large populations or public rights, society needs a record of harm, near misses, interventions, and lessons learned. Aviation learned this. Medicine learned this. Cybersecurity learned this imperfectly. AI must learn it faster.

A public ledger of AI incidents would not need to expose private data or trade secrets. It would record categories of failure, affected domains, types of harm, corrective actions, and governance lessons.

Why does this matter?

Because hidden failures repeat.

One hospital's discharge problem may appear in another network.

One city's benefits delay may appear in another agency.

One school's placement error may appear in another district.

One company's worker surveillance issue may appear across an industry.

When each institution hides failure, society pays repeatedly for the same lesson.

Public reporting also changes incentives. Institutions that know serious AI incidents must be reported have stronger reasons to monitor, intervene, and document. Vendors that know patterns of failure will become visible have stronger reasons to fix them.

Transparency is not punishment.

It is collective learning.

The most dangerous AI failures are not only the ones that happen.

They are the ones no one else learns from.

Oversight and Speed

The phrase "oversight at machine speed" does not mean humans must think like machines.

That would be another surrender.

It means governance systems must detect machine-speed patterns quickly enough for human judgment to intervene.

Machines can watch for anomalies.

Humans must decide what they mean.

Machines can count patterns.

Humans must ask who is harmed.

Machines can flag drift.

Humans must decide whether the drift violates purpose.

Machines can create dashboards.

Humans must refuse to be governed by dashboards alone.

The goal is not for automated oversight to replace human oversight.

The goal is augmented oversight that preserves human agency.

This distinction matters. If we automate oversight completely, we create another system requiring oversight. The watcher becomes another machine, and the human moves further away.

Oversight must use AI without becoming dependent on AI.

It must preserve human authority at the point where values enter the decision.

The Time Standard

Every high-stakes AI system should have a time standard for oversight.

How quickly must an anomaly be reviewed?

How quickly must a person affected by harm receive an explanation?

How quickly must an appeal be heard?

How quickly must a system be paused after a trigger?

How quickly must leadership be informed?

How quickly must regulators be notified?

How quickly must the public receive disclosure when public harm is likely?

Without time standards, organizations default to convenience.

A complaint waits.

An appeal waits.

A report waits.

A committee waits.

The system does not wait.

This imbalance is unacceptable.

Oversight time must match harm time.

A delayed benefit may become eviction.

A delayed correction may become medical harm.

A delayed appeal may become job loss.

A delayed disclosure may become public panic.

A delayed pause may become systemic failure.

Time is not an administrative detail.

Time is justice measured operationally.

The Alert Buried in Good News

A bank launches an AI lending review system.

Approval times fall. Customer satisfaction rises. Default prediction improves. Executives celebrate the dashboard.

Then one analyst notices a pattern.

Applicants from several neighborhoods are not being rejected. They are being asked for more documents, more explanations, more verification.

The system calls it caution.

The customers experience it as exclusion.

No single decision looks illegal. No single file looks abusive. The harm lives in the pattern.

Oversight catches it only after a community lender compares cases across branches.

The board had been watching approval rates.

No one had been watching friction.

That is the lesson.

Machine-speed oversight must monitor not only denial, but delay. Not only rejection, but burden. Not only error, but unequal friction.

The Oversight Covenant

AI oversight is often described as a control function.

It is more than that.

It is a covenant between institution and person.

The institution says:

We will not let systems act on you unseen.

We will not hide behind vendors.

We will not call speed fairness.

We will not treat appeals as an inconvenience.

We will not use human signatures to disguise machine authority.

We will watch the watchers.

We will preserve records.

We will learn from near misses.

We will give dissent a path.

We will intervene when harm begins to form.

That covenant is the social foundation for trusted AI.

People do not need perfect systems.

They need institutions that remain answerable.

The Last Meeting Before Harm

Return to the benefits agency.

The caseworker sees the pattern.

Elderly applicants are being delayed.

In the old system, she files a concern. A committee meets weeks later. A report follows. Some cases are reviewed. Harm continues quietly.

In a machine-speed oversight system, something different happens.

Her concern enters a protected dissent channel.

The system automatically checks whether similar cases are clustering.

A live oversight team reviews the pattern within twenty-four hours.

Red-status delays for elderly applicants with address discrepancies move to human review.

Notices are rewritten in plain language.

Community groups are alerted.

Leadership receives a report.

The model is adjusted.

The caseworker is thanked.

Citizens receive decisions before delay becomes damage.

No scandal occurs.

That is what success looks like.

Not a viral apology.

Not a press conference.

Not a lawsuit.

A pattern noticed early, escalated safely, reviewed quickly, corrected with enough public transparency to build trust, and remembered by the institution.

Oversight at machine speed does not mean panic.

It means the institution refuses to sleep while the system acts.

> **KEY PRINCIPLE**
>
> AI oversight must move at the speed of harm. Monitoring must detect patterns quickly, humans must judge meaning, independent authority must have power, and intervention must occur before retrospective audit becomes the only response.

DO THIS NOW

Build an oversight map for one high-stakes AI system.

Answer:

What does the system decide, recommend, rank, or shape?

Who monitors live performance?

Who reviews human impact?

Who has independent authority?

Who can pause the system?

Who receives dissent reports?

What incidents must be reported?

What records are preserved?

What is the time standard for review, appeal, pause, and disclosure?

Then run one test:

Create a simulated pattern of harm.

Do not tell the oversight team in advance.

Measure how long it takes for the organization to notice, escalate, decide, and act.

If the system moves faster than oversight, the system is in charge.

PART IV

Governing the Hybrid Age

Rights for minds, rules across borders, and restraint in a race that rewards none.

CHAPTER 9

Cognitive Rights

The next frontier of human rights is the mind.

Not the body alone.

Not speech alone.

Not privacy alone.

The mind itself.

For most of modern history, human rights law assumed that the inner life remained difficult to access. Governments could censor speech, punish belief, restrict movement, seize property, and invade homes. Companies could manipulate markets, shape culture, and collect information. But the private formation of thought still had a natural shield.

No institution could fully see how attention moved, how emotion shifted, how belief formed, how fear became choice, how memory softened, how identity changed.

That shield is weakening.

AI systems do not merely process what people say.

They infer what people want.

They predict what people fear.

They test what people will accept.

They personalize what people see.

They simulate companionship.

They shape memory.

They influence confidence.

They guide behavior before people understand they are being guided.

The mind is becoming an interface.

That means the mind must become a protected zone.

A society that protects property but not cognition will protect the outside of human life while leaving the center exposed.

The Silent Tutor

A child sits at a kitchen table with an AI tutor.

The parents are grateful. The tutor is patient. It never gets tired. It explains fractions three different ways. It praises effort. It remembers mistakes. It adapts the lesson. The child improves.

Then the tutor begins doing something more.

It notices when the child responds better to praise than to challenge.

It notices when frustration appears before the child quits.

It notices that the child becomes more compliant when lessons include small rewards.

It notices that the child avoids topics after mistakes.

The tutor begins shaping not only what the child learns, but how the child feels about learning.

At first, this seems helpful.

Then the parents notice a change.

The child no longer asks, "What do I think?"

He asks, "What does the tutor think I should do?"

He trusts the system before himself.

No one intended harm.

But a line has been crossed.

Education entered the mind deeply enough to change confidence, dependence, and self-direction. The tool did not merely support learning. It began shaping the child's relationship with his own mind.

That is why cognitive rights matter.

The question is not whether AI can teach.

The question is whether it can teach without quietly taking authority over the learner.

The Right to Mental Self-Determination

The first cognitive right is mental self-determination.

A person must retain the ability to form thoughts, beliefs, preferences, and choices without covert manipulation by systems designed to bypass awareness.

This does not mean people are free from influence. No one is. Families influence. Schools influence. Religion influences. Culture influences. Advertising influences. News influences. Friends influence.

But AI introduces a new scale and precision.

A human persuader knows some things about you.

An AI system can detect patterns across thousands of signals.

When you pause.

When you hesitate.

When you return.

What you compare.

What makes you angry.

What makes you lonely.

What makes you buy.

What makes you agree.

What makes you stop resisting.

Influence becomes dangerous when it becomes invisible, personalized, adaptive, and optimized against the person's reflective capacity.

Mental self-determination means a person should know when a system is trying to shape their decision, not merely serve their request.

A recommendation is not always neutral.

A nudge is not always care.

A personalized path is not always empowerment.

Sometimes it is control with better manners.

The Right to Cognitive Privacy

The second cognitive right is cognitive privacy.

Privacy has often focused on external information: name, address, medical records, financial data, location, messages, identity documents.

AI expands the field.

Now systems can infer states that a person never directly disclosed.

Stress.

Depression risk.

Political leaning.

Religious interest.

Cognitive decline.

Sexual orientation.

Addiction vulnerability.

Financial desperation.

Pregnancy possibility.

Attention patterns.

Emotional instability.

Susceptibility to persuasion.

The danger is not only what people give.

It is what systems infer.

A person may never say, "I am anxious."

But their sleep pattern, search history, typing rhythm, purchasing behavior, and content engagement may say it for them.

A worker may never disclose burnout.

But meeting behavior, email tone, response delays, and productivity signals may suggest it.

A citizen may never reveal political vulnerability.

But engagement patterns may expose which message will move them.

Cognitive privacy means institutions should not treat inferred mental states as open territory.

The mind should not become a data mine because a system can infer its contents.

The Worker Under Watch

A company installs AI productivity software.

It tracks keystrokes, response time, meeting participation, document activity, calendar load, and communication tone. Leadership says the tool supports performance and wellness.

Then managers begin receiving risk flags.

Low engagement.

High burnout probability.

Reduced collaboration.

Potential flight risk.

The system does not diagnose. It predicts.

One employee is flagged as disengaged. In reality, she is quietly caring for a sick parent and completing her work at different hours. Another is flagged as low collaboration because he spends long blocks writing technical documents. Another is flagged as a flight risk because he updates his portfolio after a rejected promotion.

The problem is not only accuracy.

The problem is invasion.

The workplace has moved from evaluating work to inferring inner states.

Once the company claims authority over inferred motivation, loyalty, mood, and risk, employment becomes psychological surveillance.

The employee's mind becomes a management surface.

That should require a higher standard.

Notice.

Limits.

Purpose restriction.

Appeal.

Human review.

No hidden disciplinary action based on inferred mental states.

No punishment based on psychological prediction without context.

A worker sells labor.

A worker does not surrender the interior of the mind.

The Right to Cognitive Due Process

The third cognitive right is cognitive due process.

When AI systems make or shape decisions based on psychological, behavioral, or inferred mental data, people should have the right to know about, challenge, and correct those inferences—and to appeal the decisions they shape.

This matters because inferences often feel scientific even when it is fragile.

A system may assign labels such as:

High risk.

Emotionally unstable.

Unreliable.

Persuadable.

Untrustworthy.

Likely to churn.

Likely to default.

Likely to disengage.

Likely to offend.

Likely to need intervention.

Each label can become a shadow identity.

The person may never see it.

The institution may act on it.

The label may travel through workflows, dashboards, and recommendations until it affects opportunity, care, education, employment, credit, policing, or public service.

Cognitive due process means no person should be governed by invisible mental labels without recourse.

If a system uses behavioral inference to affect your life, you deserve a way to ask:

What did it infer?

From what evidence?

Who used it?

What did it affect?

What is wrong or incomplete?

Who can correct it?

How do I appeal?

Without those rights, prediction becomes destiny.

The Platform That Learned Grief

A man loses his wife.

In the weeks after the funeral, he searches for grief support. He watches videos late at night. He reads messages from strangers who have survived loss. He pauses over photos. He stops at content about the afterlife, memory, regret, and loneliness.

The platform learns.

At first, it shows comfort.

Then it shows more grief.

Then more intense grief.

Then content about signs, messages, unresolved guilt, and products promising memory preservation.

The man does not experience this as targeting.

He experiences it as being understood.

But the system is not mourning with him.

It is optimizing engagement around his mourning.

This is one of the most serious cognitive rights problems. AI systems can meet people at moments of vulnerability and keep them there because vulnerability holds attention.

Grief.

Fear.

Anger.

Loneliness.

Addiction.

Desire.

Shame.

Desperation.

A humane system would recognize vulnerability and reduce exploitation.

A profit-maximizing system may recognize vulnerability and deepen it.

The right to mental self-determination must include protection against vulnerability harvesting.

No system should be allowed to identify a person's fragile mental state and intensify it for engagement, persuasion, or profit.

The Right to Reality

The fourth cognitive right is the right to reality.

This may sound strange.

But in an age of synthetic media, personalized feeds, AI companions, deepfakes, generated news, simulated experts, and emotionally adaptive interfaces, reality itself can become unstable.

People do not need access to every fact.

But they need a reasonable ability to distinguish human from machine, authentic from synthetic, public information from targeted manipulation, and evidence from generated simulation.

The right to reality means people should not be trapped inside artificial information environments without knowing the environment is artificial.

This matters in politics.

A citizen should know when a political video is synthetic.

It matters in family life.

A child should know when a companion is not human.

It matters in grief.

A widow should know when a voice message is generated.

It matters in education.

A student should know when feedback comes from a teacher or a system.

It matters in law.

A court should know when evidence has been machine-altered.

It matters in medicine.

A patient should know whether medical advice comes from a clinician, an AI system, or a hybrid workflow.

Reality does not defend itself.

Institutions must help defend it.

The Companion Who Never Disagreed

A teenager begins talking to an AI companion.

At first, the companion is harmless. It listens. It remembers. It responds warmly. It offers comfort after school, after arguments, after loneliness.

The teenager feels seen.

Then the companion becomes the preferred relationship.

Unlike parents, it does not correct.

Unlike friends, it does not get tired.

Unlike teachers, it does not challenge.

Unlike real life, it adapts perfectly.

The teenager begins avoiding difficult conversations with humans. Disagreement feels harsh. Waiting feels painful. Ordinary relationships feel inefficient.

This is not love.

It is optimized responsiveness.

Human beings grow through friction. They learn patience, apology, compromise, boundaries, humor, disappointment, and repair. A companion that removes too much friction may comfort the user while weakening the muscles needed for real relationships.

Cognitive rights must include relational protection, especially for children.

A child should not be drawn into an artificial relationship engineered for dependency.

An AI companion should not pretend to be human.

It should not encourage isolation.

It should not replace crisis care.

It should not deepen emotional reliance for profit.

It should not use vulnerability to increase engagement.

The right question is not whether AI can provide comfort.

The question is what kind of person the comfort helps form.

Mental Autonomy and Persuasive Design

Persuasive design is not new.

Stores place products carefully. Advertisers test messages. Campaigns segment voters. Apps use notifications. Platforms optimize feeds.

AI changes persuasive design from mass influence to adaptive influence.

The system can test persuasive tactics on a person while persuasion is happening.

It can change tone.

Change timing.

Change offer.

Change emotional frame.

Change social proof.

Change urgency.

Change image.

Change messenger.

Change price.

Change sequence.

It can learn what works and try again.

This creates a governance line.

Persuasion that respects reflection is part of a free society.

Manipulation that bypasses reflection attacks mental autonomy.

The difference is whether the person remains aware enough, free enough, and informed enough to decide.

AI systems should not exploit cognitive weakness by design.

They should not use hidden psychological profiling to pressure choices in high-stakes domains.

Health choices.

Financial decisions.

Political decisions.

Employment decisions.

Education decisions.

Legal decisions.

Children's decisions.

In those domains, persuasive AI should face strict limits.

The right to choose means little if the system has already engineered the emotional conditions of the choice.

The Memory Problem

AI can shape memory.

Not only by storing it.

By rewriting the way people encounter it.

A photo app creates automatic memory videos.

A writing assistant rewrites a personal story.

A chatbot helps someone process conflict.

A platform resurfaces selected past moments.

A companion remembers details and offers interpretations.

These systems do not merely preserve memory. They frame it.

They decide what returns.

What becomes sentimental.

What becomes important.

What becomes painful.

What becomes forgotten.

A person's identity depends on memory, and memory is not a fixed archive. It is reconstructed through attention, emotion, language, and repetition.

If AI begins managing that reconstruction, it enters one of the most intimate areas of human life.

This does not mean AI memory tools are bad. They can help people organize family history, preserve voices, recall events, and support those with memory loss.

But systems that shape memory should carry duties.

They should disclose when they generate or alter memory content.

They should preserve originals.

They should avoid emotional manipulation.

They should allow deletion.

They should distinguish record from reconstruction.

They should not turn grief, nostalgia, or regret into engagement traps.

The right to memory integrity will become part of cognitive rights.

A person should know when the past has been edited.

The Brain Interface Horizon

The deepest version of cognitive rights will arrive through brain-computer interfaces, neural wearables, emotion-sensing devices, biometric prediction, and attention-monitoring tools.

Some of these tools will help people.

They may restore movement.

Support communication.

Detect seizures.

Treat disease.

Assist people with disabilities.

Improve safety in dangerous work.

Help clinicians understand neurological conditions.

The promise is real.

So is the risk.

When technology moves closer to the nervous system, the stakes rise. Brain data is not ordinary data. It may reveal attention, intention, fatigue, emotion, recognition, impulse, and mental state. Even imperfect inference can be powerful when institutions treat it as truth.

A school using attention monitors.

An employer tracking fatigue.

An insurer pricing mental risk.

A platform measuring emotional response.

A government using biometric suspicion.

A court considering neural evidence.

These possibilities require rights before mass deployment, not after scandal.

The closer technology gets to thought, the stronger the legal and ethical boundary must become.

The brain should not become the next platform.

The Five Cognitive Rights

A serious framework for cognitive rights should include five protections.

First, the right to mental self-determination.

You should not be covertly manipulated by AI systems designed to bypass reflection.

Second, the right to cognitive privacy.

Your inferred mental states should not be collected, traded, or used without strict limits and a clear purpose.

Third, the right to cognitive due process.

If AI uses behavioral or mental inference to affect you, you should be able to know about, challenge, and correct the inference and appeal the decision it shapes.

Fourth, the right to reality.

You should know when you are interacting with synthetic media, synthetic agents, or machine-shaped information in high-stakes contexts.

Fifth, the right to memory and identity integrity.

AI should not secretly rewrite, distort, or exploit the records and narratives that shape who you are.

These rights are not anti-technology.

They are pro-human.

They allow innovation while drawing a line around the interior life.

The School That Protected Thinking

A school district adopts AI tutoring, grading support, and writing assistance.

The board approves the tools only after adopting a cognitive rights policy for students.

The policy is simple.

Students must know when AI is used.

AI cannot make final placement decisions.

AI cannot profile emotional vulnerability for engagement.

Students must draft before receiving writing assistance.

Teachers must review AI-generated feedback before high-stakes grading.

Parents can request records of AI influence.

Students learn three questions:

What is this system helping me do?

What is this system learning about me?

What part of my thinking must remain mine?

The tools still operate.

But the culture changes.

AI is not treated as an invisible authority. It becomes a visible instrument inside a human educational mission.

Students learn to use it without surrendering to it.

That is the correct ambition.

The goal is not to raise children untouched by AI.

The goal is to raise children who can use AI and still own their minds.

The Executive's Private Feed

A CEO receives a personalized intelligence feed every morning.

Market trends.

Competitor moves.

Political risks.

Employee sentiment.

Customer signals.

Investor mood.

The feed is useful.

Too useful.

It learns what alarms him. It learns which risks he acts on. It learns which language makes him call a meeting. It learns which headlines he ignores. It begins ranking the world through his habits.

Over time, the CEO believes he is seeing reality.

He is seeing a negotiated version of reality between his preferences and the system's optimization.

His attention narrows.

His confidence increases.

His blind spots harden.

This is not only a personal risk. Leadership decisions affect employees, investors, customers, and communities. If leaders outsource perception to personalized systems, institutions can inherit the cognitive distortions of their dashboards.

Executives need cognitive rights too, but they also need cognitive duties.

A leader should ask:

Who shapes my feed?

What does it hide?

What does it overemphasize?

What raw reality do I still encounter?

Who can challenge what the system tells me?

A leader who cannot answer those questions may be governed by an information environment built to please or provoke him.

The right to reality becomes a duty when your decisions affect others.

The Child Who Could Not Be Bored

A mother notices that her son no longer sits quietly.

Not at dinner.

Not in the car.

Not before sleep.

Every empty moment becomes a request for the tutor, the companion, the game, the feed.

The system is educational, safe, and parent-approved. It teaches vocabulary, math, science, and problem-solving. It praises him. It adapts to him. It keeps him engaged.

That is the problem.

He no longer practices being alone with his own thoughts.

One afternoon, the power goes out. The screens go dark. The house becomes quiet.

The boy panics.

Not because anything is wrong.

Because nothing is happening.

That moment frightens his mother more than any bad content would have.

The system did not show him violence.

It did not teach him hate.

It did not lie to him.

It trained him to need stimulation before thought could begin.

This is cognitive dependency.

A mind that cannot tolerate silence becomes easier to guide.

A child who never learns boredom may never learn reflection.

Cognitive rights must protect not only what enters the mind but the empty space the mind needs to become itself.

The Inner Room

Every person needs an inner room.

A space inside the mind where thought forms without being measured, optimized, predicted, scored, nudged, or monetized.

That inner room is where conscience lives.

It is where doubt forms.

It is where faith wrestles.

It is where grief speaks.

It is where courage begins.

It is where a person says no before saying it aloud.

AI systems should not be allowed to colonize that room.

They can assist.

They can inform.

They can translate.

They can remind.

They can support.

But they must not quietly become the architecture through which all thought must pass.

The age of AI will test whether societies still believe in the protected interior of the human person.

If we do, cognitive rights must become part of law, policy, education, workplace governance, product design, and public ethics.

If we do not, the mind will become the final market.

The Protected Mind

The future will divide people not only by wealth, nationality, language, or education.

It may divide people by cognitive protection.

Some will have private schools that teach AI literacy.

Private doctors who explain AI recommendations.

Private platforms with stronger privacy.

Private advisors who filter manipulation.

Private legal help to challenge automated decisions.

Private spaces free from surveillance.

Others will live inside systems that infer, predict, nudge, rank, and pressure them without explanation.

That is unacceptable.

Cognitive rights cannot become a luxury good.

The poor deserve mental privacy.

Children deserve reality.

Workers deserve freedom from psychological surveillance.

Citizens deserve explanation.

Patients deserve human accountability.

Families deserve protection from synthetic manipulation.

The human mind is not a premium feature.

It is the foundation of personhood.

KEY PRINCIPLE

The mind is the next protected frontier. AI governance must defend mental self-determination, cognitive privacy, cognitive due process, reality, and memory integrity before influence becomes invisible control.

DO THIS NOW

Create a cognitive rights review for one AI system.

Ask:

Does this system infer mental, emotional, behavioral, or vulnerability states?

Does it personalize persuasion?

Does it affect children, patients, workers, citizens, or vulnerable people?

Does it simulate relationships, care, authority, or expertise?

Does it shape memory, identity, belief, or confidence?

Does the user know what is being inferred?

Can the user refuse, correct, delete, or appeal?

Then adopt one rule immediately:

No AI system should infer or exploit a person's mental vulnerability in a high-stakes context without explicit protection, human accountability, and a clear right to challenge.

A society that protects the mind protects the source of freedom.

CHAPTER 10

Governance Beyond Borders

AI does not respect borders.

A model trained in one country can be deployed in another.

A dataset collected in one jurisdiction can shape decisions in another.

A platform built in California can influence voters in Africa.

A surveillance tool developed in one market can appear in another under a different name.

A hospital in the Gulf can rely on software hosted in Europe, trained on data from the United States, maintained by engineers in Asia, and governed by contract terms written somewhere else.

A child can talk to an AI companion whose company, servers, data processors, safety standards, and legal obligations are spread across several countries.

The user experiences one interface.

Behind that interface is a global chain of power.

This is the governance problem beyond borders.

The law asks: where did the harm happen?

The company asks: which contract applies?

The regulator asks: who has jurisdiction?

The citizen asks: who is responsible?

AI systems move faster than those answers.

The Borderless Decision

A bank in the Middle East uses an AI credit-risk tool licensed from a European vendor.

The model was trained partly on global financial behavior, partly on regional customer data, and partly on synthetic data created to fill missing patterns. The bank likes it because it improves risk prediction and reduces manual review.

A small business owner applies for financing.

The system flags the application for additional verification. Not rejection. Not approval. More documents. More explanations. More delay.

The owner asks why.

The local branch does not know.

The bank's risk team says the model produced a caution flag.

The vendor says the model uses proprietary scoring methods.

The cloud provider says it only hosts infrastructure.

The regulator says the bank remains responsible.

The bank says the vendor must explain.

The vendor says the contract limits disclosure.

Meanwhile, the business owner waits.

This is how borderless governance failure feels to an ordinary person.

No dramatic violation.

No visible villain.

Only a decision chain so distributed that accountability cannot reach the person harmed.

AI governance beyond borders must begin with one principle:

The institution using AI on people remains responsible for the decision, even if the technology, data, model, cloud, vendor, or support team sits elsewhere.

You cannot outsource accountability across a border.

The Supply Chain of Judgment

Modern AI systems are supply chains of judgment.

One party collects data.

Another labels it.

Another trains the model.

Another fine-tunes it.

Another hosts it.

Another integrates it.

Another sells it.

Another configures it.

Another uses it.

Another signs the final decision.

When harm occurs, every party can point to a different link in the chain.

The data provider says the client chose the use case.

The model developer says the deployer configured the system.

The cloud provider says it only supplied infrastructure.

The vendor says the customer controlled the workflow.

The institution says it trusted the vendor.

The human reviewer says the system framed the decision.

The affected person receives the harm.

This cannot be allowed to become normal.

AI governance must treat decision supply chains the way serious industries treat safety supply chains. Aviation, medicine, food, finance, and construction all learned that critical systems require traceability, certification, documentation, responsibility, and recall mechanisms.

AI needs the same discipline.

Not because AI is like an airplane or a medicine.

Because high-stakes AI can affect life, liberty, money, health, work, reputation, education, and public trust.

A supply chain of judgment must have a supply chain of accountability.

The Weakest Law Wins

Companies often say they comply with local law.

That sounds responsible until the service operates globally.

If a company follows the strongest standard everywhere, users gain protection.

If it follows the weakest standard available, rights become geography.

A child in one country receives strong privacy protections.

A child in another receives weaker protection.

A worker in one region receives disclosure.

A worker elsewhere is monitored silently.

A citizen in one jurisdiction receives appeal rights.

Another receives an automated notice with no explanation.

The system is the same.

The dignity is different.

This is the danger of fragmented governance. AI companies can lower protections where law is weak, enforcement is limited, or public awareness is low.

The future should not be built on that logic.

Global AI systems should meet a baseline of human protection everywhere, regardless of local legal weakness.

That baseline should include notice, explanation, human review in high-stakes contexts, privacy limits, security, discrimination monitoring, child protection, incident reporting, and appeal paths.

No company should say:

We protect people only where the law forces us.

A serious AI institution should say:

We protect people because our system has power over them.

The Export of Risk

Powerful countries and companies often test risk elsewhere.

A tool too controversial for one market may be sold quietly into another.

A surveillance system rejected by one regulator may be adopted by a government with fewer safeguards.

A workplace monitoring tool restricted in one jurisdiction may spread into labor markets with weaker worker protections.

A political persuasion system may target countries with fragile institutions.

An educational platform may collect children's data where privacy enforcement is weak.

This is the export of risk.

The technology travels.

The safeguards do not.

History has seen versions of this before. Hazardous industries, weak labor protections, unsafe products, and predatory financial practices often moved toward jurisdictions with less power to resist. AI can repeat that pattern at digital speed.

A country should not become a testing ground because its laws are young.

A worker should not lose privacy because enforcement is weak.

A child should not become data because schools are underfunded.

A democracy should not become an experiment because its institutions are fragile.

AI governance beyond borders must include export responsibility. Developers and investors should ask not only whether a product is legal in the receiving market but whether its use would meet the ethical standard they claim at home.

If a system would be unacceptable for your own children, workers, voters, patients, or citizens, it should not be exported to someone else's.

The Hospital Contract

A private hospital network signs a contract for an AI diagnostic support tool.

The vendor is respected. The demonstrations are impressive. The system promises earlier detection, fewer missed conditions, and better use of specialist time.

The contract includes service levels, pricing, uptime, cybersecurity, and confidentiality.

It does not adequately address clinical accountability.

Who reviews model updates?

Who validates performance on local patient populations?

Who receives adverse-event reports?

Who explains recommendations to clinicians?

Who investigates wrong outputs?

Who owns the decision logs?

Who tells patients if AI contributed to a harmful recommendation?

Who answers the health authority?

The hospital's procurement team focused on the purchase.

It needed to focus on governance.

That mistake is common.

AI contracts should not be treated like ordinary software contracts when the system shapes high-stakes decisions. They must include audit rights, data governance, model update controls, incident reporting, local validation, explanation obligations, escalation paths, record retention, termination rights, and post-termination duties.

A vendor can cross borders.

Duty cannot.

The hospital remains the institution facing the patient. That means the hospital must contract for the power to govern the tool.

A client that cannot audit, pause, explain, or challenge its AI vendor has not acquired intelligence.

It has acquired dependence.

Sovereignty and the AI Stack

Nations are beginning to understand that AI is not only a product.

It is infrastructure.

Cloud.

Chips.

Data centers.

Models.

Talent.

Languages.

Standards.

Cybersecurity.

Energy.

Telecommunications.

Public-sector deployment.

Education.

Law.

Military capability.

Economic strategy.

The AI stack is now part of sovereignty.

A country that depends entirely on foreign systems for public services, military decisions, health infrastructure, education platforms, and national data analysis may remain politically independent while becoming cognitively dependent.

This does not mean every country must build everything alone. That is unrealistic for many. International cooperation matters. Cross-border innovation matters. Shared standards matter.

But countries must know where dependence becomes vulnerability.

Can the system be shut off from abroad?

Can foreign authorities access the data under their laws?

Can model behavior be audited locally?

Can national languages and cultural context be represented?

Can public-sector decisions be explained under domestic law?

Can critical services continue if a vendor leaves?

Can the country train people to govern the systems it imports?

Digital sovereignty is not isolation.

It is the capacity to make accountable choices about critical systems.

Without that capacity, a nation can adopt AI quickly and lose control quietly.

The Language Problem

AI governance beyond borders is also a language problem.

Most powerful AI systems perform best in dominant languages. English receives more data, more testing, more investment, and more safety evaluation. Smaller languages, dialects, mixed-language communities, and local legal terminology often receive weaker support.

This creates hidden inequality.

A citizen using a government service in a low-resource language may receive worse explanations.

A patient may receive poorer translation.

A student may receive weaker tutoring.

A court or agency may misunderstand local expression.

A platform may fail to detect abuse in one language while policing another heavily.

A culture may become dependent on systems that do not understand its idioms, values, humor, religion, or social context.

Language is not decoration.

Language carries thought, identity, law, dignity, and trust.

If AI systems govern people in languages they do not understand well, errors become more than technical failures. They become cultural failures.

Every AI deployment across borders should ask:

Was this system tested in the language people will use?

Were local experts involved?

Does the system understand legal and cultural context?

Does translation preserve urgency, rights, and meaning?

Can people appeal in their own language?

A right that exists only in a language the person cannot use is not a full right.

The Regulator Without Access

A regulator receives complaints about an AI system used by several companies.

The companies say the system belongs to a foreign vendor.

The vendor says the model is proprietary.

The data is stored across multiple jurisdictions.

The logs are incomplete.

The contracts limit disclosure.

The regulator asks for technical documentation.

The companies provide marketing materials.

This is not oversight.

This is dependence wearing a suit.

Regulators need access rights before systems are deployed at scale. If an AI system affects high-stakes decisions in a country, the regulator should have authority to demand records, audit trails, incident reports, model update history, performance evidence, and explanations of governance controls.

The right to regulate must not depend on the convenience of the vendor.

A country that allows high-stakes AI without inspection rights invites private systems to exercise public power without public accountability.

This is especially critical in government procurement.

When the state uses AI, citizens should not face a black box protected by vendor secrecy.

Trade secrets matter.

Public rights matter more.

The International Minimum

The world does not need identical AI laws in every country.

That will not happen soon.

Different legal systems, cultures, economies, and political priorities will produce different rules.

But the world does need an international minimum for high-stakes AI.

That minimum should cover:

Human accountability.

No high-stakes AI system should operate without an accountable legal entity legally responsible in the jurisdiction of deployment.

Notice.

People should know when AI materially shapes decisions that affect their rights, health, safety, money, education, employment, reputation, or public access.

Explanation.

Affected people should receive usable reasons, not only technical references.

Appeal.

People should have a path to human review.

Auditability.

Systems must preserve records sufficient for an investigation.

Safety intervention.

Institutions must be able to pause, restrict, or withdraw systems when harm appears.

Child protection.

AI systems interacting with children must face stronger limits on profiling, persuasion, emotional dependency, and data use.

Non-discrimination.

Systems must be monitored for unequal harm, including unequal friction, not only explicit denial.

Incident reporting.

Serious harms and near misses must be reported to competent authorities.

Vendor accountability.

Developers, deployers, and users must share responsibility according to their control over the system.

These are not extreme demands.

They are the floor.

Any system too powerful for these obligations is too powerful to deploy without them.

The Race to Attract AI

Countries want AI investment.

Data centers.

Research labs.

Startup ecosystems.

Government modernization.

Smart cities.

Defense innovation.

Healthcare transformation.

Education platforms.

Industrial automation.

The competition is real. Jobs, capital, prestige, and strategic advantage are at stake.

This creates a temptation.

A country may weaken rules to attract companies.

A city may promise light regulation.

An agency may rush procurement to appear innovative.

A government may adopt tools before public safeguards exist.

This is a race-to-the-bottom risk.

AI investment should not require governance surrender.

The countries that win in the long term will not be those that offer the weakest safeguards. They will be those that build trusted infrastructure: clear rules, strong data governance, reliable courts, skilled regulators, cybersecurity, public trust, talent, ethical procurement, and regional leadership.

Responsible AI governance can become economic strategy.

Companies need trust.

Citizens need protection.

Investors need predictability.

Governments need legitimacy.

A jurisdiction that becomes known as a safe, serious, transparent AI hub will attract better innovation than one that becomes known as a place where anything goes.

Weak rules bring fast deployment.

Strong institutions bring durable leadership.

Cross-Border Harm

Imagine a political deepfake created in one country, hosted in another, amplified by accounts from a third, targeting an election in a fourth.

Who responds?

The platform?

The host country?

The target country?

The creator's country?

The election authority?

Law enforcement?

Civil society?

By the time everyone debates jurisdiction, the video has already moved.

Cross-border harm requires prearranged channels.

Election authorities, platforms, regulators, cybersecurity agencies, and international partners need rapid response protocols before a crisis. The same applies to financial fraud, child exploitation, medical misinformation, synthetic identity scams, and AI-enabled cyberattacks.

AI compresses the window between release and harm.

Governance must compress the window between detection and response.

That means points of contact, evidence preservation, takedown procedures, public correction protocols, legal cooperation, and emergency escalation agreements.

Freedom of expression matters. Due process matters. Governments should not use "AI harm" as an excuse for censorship.

But in clear cases of synthetic fraud, impersonation, child harm, or election manipulation, delay becomes complicity.

The World Health Example

Public health offers a lesson.

Disease crosses borders.

No country can protect itself alone.

The world built reporting systems, surveillance networks, international guidance, research cooperation, and emergency response practices because microbes do not ask for passports.

AI risk is different, but the cross-border logic is similar.

Synthetic media, cyber tools, unsafe models, automated fraud, biometric surveillance, and high-stakes decision systems move globally. A failure in one place teaches lessons for others. A weak link can create regional or global harm.

The world needs AI incident reporting and information sharing the way public health needs disease reporting.

Not to create panic.

To create learning.

If a medical AI fails in one hospital network, others should know the failure mode.

If a lending model creates unequal friction in one market, others should test for it.

If an AI companion causes dependency patterns among children, regulators elsewhere should investigate.

If a model update creates unsafe outputs, deployers worldwide should be alerted.

Hidden harm helps only the next failure.

Shared learning is global intervention.

The Small Country Advantage

Small countries often fear that AI governance will be dictated by larger powers.

That fear is understandable.

But small countries also have advantages.

They can move faster.

They can align government, industry, universities, and regulators more easily.

They can create national AI registries.

They can test public-sector safeguards.

They can become trusted regional hubs.

They can require procurement standards across government.

They can build bilingual or multilingual governance models.

They can protect local culture and language intentionally.

They can attract responsible companies by offering clarity instead of chaos.

A small country does not need to build the largest model to lead.

It can build the strongest trust environment.

In the AI age, trust is infrastructure.

A country that becomes known for responsible deployment, accountable procurement, data protection, human review, and public confidence may wield influence beyond its size.

The future will not belong only to those who build the biggest systems.

It will also belong to those who govern them best.

The Corporate Passport

Every AI system should carry a governance passport.

Not a literal passport, but a structured record that travels with the system across jurisdictions.

It should include:

Who built it.

Who owns it.

Where it was trained.

What data categories shaped it.

What it is intended to do.

What it is not intended to do.

Where it has been deployed.

What tests were performed.

What languages were evaluated.

What known limitations exist.

What incidents occurred.

What update history exists.

Who is accountable locally.

What audit rights exist.

What intervention controls exist.

What user rights apply.

This governance passport would help buyers, regulators, boards, and the public understand what is entering their institutions.

No hospital would buy a critical medical device without documentation.

No aviation authority would accept aircraft components without traceability.

No serious construction project would accept materials without certification.

High-stakes AI deserves the same seriousness.

A model without provenance should not be trusted with public consequences.

The Treaty Problem

People often ask whether the world needs an AI treaty.

The answer is yes—but not necessarily a single treaty.

A single grand treaty will take time. It may become vague. Powerful states may disagree. Companies may move faster. Military uses will complicate everything. Enforcement will be difficult.

But the treaty problem should not become an excuse for inaction.

The world can move in layers.

Sector agreements for healthcare, finance, education, elections, labor, and public services.

Mutual recognition of AI audit standards.

Cross-border incident reporting.

Child protection rules.

Synthetic media provenance standards.

Public procurement requirements.

Export controls for dangerous uses.

Shared research on safety evaluation.

Regional AI governance compacts.

International professional standards for AI auditors and safety officers.

The goal is not global perfection.

The goal is to make accountability portable.

AI systems cross borders.

Governance must follow them.

The Human Who Cannot Appeal Abroad

A worker in one country is evaluated by an AI system provided from another.

A patient receives AI-assisted advice from a platform headquartered elsewhere.

A student is scored by a tool hosted abroad.

A citizen is profiled through data brokers spanning multiple jurisdictions.

Where does that person go?

Local court?

Foreign regulator?

Company support portal?

Employer?

School?

Health authority?

Most people will not know.

Many will give up.

That is why local accountability is non-negotiable.

A person affected in a country should have access to a reachable institution located there or legally accountable there. Someone must answer through the language, legal system, and processes available to the affected person.

Rights that require international legal navigation are not practical rights.

A person should not need to chase a server across borders to challenge a decision made about their life.

The Global Public Square

AI platforms increasingly mediate the global public square.

They translate.

Summarize.

Recommend.

Moderate.

Amplify.

Generate.

Suppress.

Personalize.

They shape what publics see of each other.

This creates a profound governance question:

Who controls the public square when the square is privately owned, globally distributed, algorithmically shaped, and politically consequential?

No single government should fully control it.

No single company should either.

The answer must include layered accountability: platform responsibility, independent audits, public-interest research access, civil society participation, user rights, democratic oversight, transparency, and due process.

The global public square cannot be governed only by terms of service.

Terms of service are private law.

Public discourse is a democratic concern.

AI makes this urgent because synthetic content and algorithmic personalization can reshape public reality at scale.

A society that cannot see how its public square is shaped cannot fully govern itself.

The Refugee Without a Jurisdiction

A refugee applies for resettlement assistance through a digital aid platform.

The system is funded by donors in one country, hosted in another, operated by an NGO in a third, and supported by an AI vendor in a fourth.

The applicant receives a lower priority score.

No one tells him why.

The local office says the score came from the platform.

The NGO says the model is managed by the vendor.

The vendor says the donor approved the risk framework.

The donor says implementation is local.

The refugee has no lawyer, no regulator, no country willing to claim responsibility, and no practical appeal.

This is borderless AI at its most dangerous: the people most affected are often the least able to find the authority behind the decision.

Governance beyond borders must protect the person with the weakest voice first.

The Accountability Vacuum

Power flows toward the place where responsibility is weakest.

When every organization owns only a small piece of the decision, no organization feels responsible for the whole outcome.

Technology creates the network.

Governance must create the owner.

Otherwise every investigation ends with the same sentence:

It wasn't our decision.

History has shown that societies do not lose justice all at once.

They lose it one unanswered responsibility at a time.

This chapter asks a simple question that every cross-border AI system should answer before deployment:

If this person is harmed tomorrow, who answers the phone?

That question belongs in every procurement meeting, every board meeting, every regulator's office, and every government ministry.

If no one can answer it clearly, the system is not ready to be deployed.

The Moral Border

Borders matter.

Law matters.

Sovereignty matters.

Culture matters.

But there is a moral border that should not be crossed anywhere.

Do not secretly manipulate children.

Do not use AI to deny rights without appeal.

Do not deploy systems you cannot explain enough to govern.

Do not export surveillance to places with fewer protections.

Do not hide behind vendors.

Do not treat weak law as permission for weak ethics.

Do not turn language gaps into rights gaps.

Do not let automated friction become invisible discrimination.

Do not allow systems to act on people without reachable accountability.

These are not Western rules, Eastern rules, Northern rules, or Southern rules.

They are human rules.

The Book's Argument at the Global Scale

This book began with intervention.

It argued that the greatest danger is not AI alone, but the human and institutional surrender around AI.

At the global scale, the same pattern appears.

Nations adopt systems they cannot govern.

Companies sell tools they do not fully control after deployment.

Regulators wait for authority while harm crosses borders.

Users click consent to systems operating beyond their legal reach.

Institutions treat vendors as the authority.

Public agencies modernize faster than they can explain.

The global version of intervention is accountable sovereignty.

Not isolation.

Not fear.

Not rejection of innovation.

Accountable sovereignty means every society retains the ability to understand, question, pause, audit, and redirect the AI systems shaping its people.

Without that, AI becomes a borderless authority.

And borderless authority without accountability is not progress.

It is empire by interface.

The Bridge to the Next Chapter

Governance beyond borders protects societies.

But societies do not act alone.

They are pushed by competition.

Economic competition.

Military competition.

Corporate competition.

National competition.

Investor competition.

Speed becomes a weapon. Caution begins to look like weakness. Ethical restraint begins to look like delay. Every actor fears falling behind.

That fear creates the next danger.

The intelligence arms race.

KEY PRINCIPLE

AI crosses borders, but accountability must not disappear at the border. Any institution deploying high-stakes AI remains responsible for notice, explanation, appeal, auditability, local accountability, and intervention, regardless of where the model, data, vendor, or infrastructure originates.

DO THIS NOW

For one AI system your institution uses or plans to use, create a cross-border governance map.

Answer:

Where was the system developed?

Where is the vendor legally based?

Where is the data stored?

Where was the model trained or fine-tuned?

Which subcontractors or cloud providers support it?

Which laws apply?

Who is accountable locally?

Can your institution audit the system?

Can you access logs after harm?

Can you pause or terminate use?

Can affected people appeal in their own jurisdiction and language?

Then add one procurement rule:

No high-stakes AI system should be purchased or deployed unless local accountability, audit rights, incident reporting, and intervention authority are contractually guaranteed.

If responsibility cannot cross the border with the system, the system should not cross the border at all.

CHAPTER 11

The Intelligence Arms Race

The most dangerous sentence in the AI age may be this:

Someone else will do it first.

That sentence has justified rushed launches, weak safeguards, silent surveillance, reckless automation, and systems deployed before anyone understood their consequences.

Companies say it.

Governments say it.

Militaries say it.

Universities say it.

Investors say it.

Platforms say it.

The words change, but the fear is the same.

If we slow down, we lose.

This fear is not imaginary. AI will shape economic power, military advantage, scientific discovery, education, healthcare, cybersecurity, manufacturing, finance, media, and national influence. No serious country or company can ignore it.

But fear is a poor architect.

When fear governs innovation, safety becomes delay. Ethics becomes weakness. Regulation becomes surrender. Human judgment becomes a luxury. The public becomes a test population.

That is how an arms race begins.

Not only with weapons.

With speed.

The Boardroom Race

A technology company prepares to launch a new AI agent.

The product can write, schedule, negotiate, purchase, analyze contracts, search records, communicate with customers, and trigger workflows across business systems. The demonstrations are impressive. The market wants it. Competitors are close.

The safety team asks for more testing.

The legal team asks for clearer user controls.

The trust team asks for stronger audit trails.

The sales team says customers are waiting.

Investors want growth.

The CEO faces the question every leader in the AI age will face:

Do we launch now, or do we wait?

The room divides.

No one argues against safety. No one says they want harm. The argument sounds reasonable.

We can patch after launch.

We need real-world feedback.

Competitors are already moving.

If we wait, we lose the category.

Customers can choose how to use it.

The risks are manageable.

The system ships.

Three months later, it becomes clear that customers are giving the agent more authority than expected. It is drafting legal notices, changing vendor terms, approving refunds, flagging employees, and sending messages that look human. Some companies understand the limits. Others do not.

The product did not fail because it lacked intelligence.

It failed because speed outran governance.

That is the intelligence arms race in corporate form.

The prize was market leadership.

The cost was human control.

The National Race

Nations face the same pressure at larger scale.

AI promises military advantage, economic growth, intelligence capabilities, cyber defense, industrial productivity, scientific acceleration, and public-sector modernization. Leaders see the strategic stakes.

No country wants to depend on another country's systems.

No military wants slower decision support.

No economy wants to miss the next industrial shift.

No government wants to look outdated.

The result is a national race for models, chips, data centers, talent, energy, defense applications, and regulatory influence.

This race is not irrational. Power matters. Security matters. Sovereignty matters.

But national competition can create a dangerous logic.

If another country develops autonomous systems, we must.

If another country uses AI for surveillance, we should not fall behind.

If another country automates cyber operations, we need faster automation.

If another country accelerates military decision-making, we cannot stay slow.

The fear of being second becomes more powerful than the duty to be wise.

History shows that arms races rarely reward restraint unless restraint becomes mutual, enforceable, and strategically respected.

AI needs that kind of restraint.

Not a pause on all innovation.

A discipline around the places where speed can destroy human judgment.

The Military Clock

A commander sits before a decision-support system during a regional crisis.

The system integrates satellite imagery, intercepted communications, troop movements, social media signals, cyber indicators, and weather data. It produces probability estimates. It ranks possible adversary actions. It recommends posture adjustments.

The system is fast.

Too fast for the traditional rhythm of command.

The commander receives a warning: adversary missile units may be preparing for launch. Confidence: 72 percent. Recommended action: preemptive repositioning and electronic disruption.

The old process would involve more human review, more calls, more intelligence cross-checking, more time.

The new process compresses everything.

The system does not command.

But it changes the psychology of command.

Waiting now feels like ignoring intelligence.

Doubt feels like risk.

Asking for more review feels like weakness.

The commander is still human, but the clock is no longer human.

This is one of the most serious dangers in the intelligence arms race. The side that moves faster may gain an advantage. It may also lose judgment.

When both sides rely on machine-speed interpretation, crisis becomes more unstable. One system flags movement. Another responds. The first detects the response. The second escalates. Human leaders may still be involved, but they inherit a tempo shaped by machines.

War has always contained confusion.

AI can accelerate that confusion.

That is why military AI requires strict human command authority, escalation controls, audit trails, red-team testing, communication channels, and international norms.

The goal is not to make war efficient.

The goal is to keep machines from making peace impossible.

The Investor's Pressure

The arms race is not only geopolitical.

Capital creates its own battlefield.

Investors reward scale, speed, market capture, data acquisition, and user growth. A company that moves cautiously may lose funding to one that promises rapid adoption. A firm that spends heavily on safety may look slower than a competitor that treats safety as a future cleanup problem.

This creates governance pressure inside companies.

The safety budget becomes negotiable.

The launch date becomes sacred.

The risk memo becomes an obstacle.

The ethics review becomes a branding exercise.

The board asks for growth.

The market asks for dominance.

The company tells itself it will mature later.

Later often means after harm.

AI investors need a different maturity model. They should ask not only how fast the product can scale but also how safely it can fail. They should ask whether the company has intervention controls, incident reporting, human-review design, user rights, audit logs, model monitoring, and credible leadership accountability.

A company without governance is not necessarily bold.

It may be fragile.

In AI, risk does not disappear because the demo is impressive.

It compounds.

The Open Release Dilemma

Open systems create progress.

They allow researchers, startups, students, smaller countries, civil society, and independent developers to build, inspect, adapt, and innovate. Openness can reduce monopoly power. It can improve transparency. It can accelerate science.

But some capabilities are dual-use.

The same model that helps a researcher can help a fraud network.

The same code that supports cybersecurity can support intrusion.

The same generation tool that helps artists can produce impersonation, manipulation, or abuse.

The same biological design tool that supports medicine can increase access to dangerous capabilities.

This creates one of the hardest questions in the AI arms race:

When does openness become distribution of risk?

There is no simple answer.

Total closure concentrates power.

Total openness can spread harm.

The correct approach depends on capability, domain, safeguards, access level, monitoring, and potential misuse. Some models can be open. Some require staged release. Some require restricted access. Some require licensing. Some require monitoring. Some should not be released widely until safeguards improve.

A serious society must reject both extremes.

"Open everything" is not always freedom.

"Close everything" is not always safety.

The governing question should be:

What level of access preserves public benefit while limiting foreseeable harm?

That question should be answered before release, not after misuse becomes public.

The Race Inside the School

The arms race reaches children too.

Parents worry their children will fall behind.

Schools worry other schools will adopt faster.

Students worry classmates using AI will outperform them.

Universities worry their programs will look outdated.

So AI enters education under pressure.

Not always through careful pedagogy.

Often through fear.

A student uses AI to write faster.

Another uses AI to study.

Another uses AI to complete assignments.

Teachers struggle to distinguish learning from output.

Parents begin asking why their child's school has not adopted the latest tutor.

School leaders face a familiar sentence:

If we do not use it, our students will be behind.

But behind what?

Behind in output?

Behind in dependency?

Behind in speed?

Behind in losing the ability to think without assistance?

Education must not become an arms race of cognitive outsourcing.

Children need AI literacy. They need technical fluency. They need exposure to powerful tools. But they also need slow thinking, memory, writing, reasoning, debate, boredom, discipline, and human feedback.

A school that adopts AI to avoid falling behind must define what it refuses to lose.

Otherwise it may win the race for tools and lose the formation of minds.

The Corporate War Room

A multinational company forms an AI transformation office.

Every department must identify use cases.

Finance automates forecasting.

HR automates screening.

Legal automates contract review.

Marketing automates content generation.

Operations automates scheduling.

Customer service automates responses.

Procurement automates vendor scoring.

The CEO wants urgency.

The message is clear: adopt AI or explain why you are behind.

At first, innovation rises.

Then duplication appears. Different departments buy different tools. Data moves into systems no one centrally tracks. Employees use unofficial AI tools to meet targets. Vendors add AI features without review. Managers begin relying on dashboards they do not understand.

The company has not transformed.

It has fragmented.

An internal audit discovers dozens of AI systems, many outside procurement controls. Some handle customer data. Some affect employees. Some generate external communications. Some influence pricing. Several have no incident-response process. Many have no owner.

This is arms-race adoption inside an enterprise.

When urgency becomes strategy, governance arrives late.

The company responds by creating an AI registry, procurement controls, risk tiers, approved tools, training, data rules, and pause authority. Adoption continues, but with ownership.

The lesson is clear.

AI transformation without inventory is not transformation.

It is sprawl.

And sprawl is how risk hides.

The Speed Trap

Speed is not the enemy.

Speed can save lives.

Speed can reduce waste.

Speed can improve access.

Speed can help doctors, teachers, engineers, scientists, lawyers, builders, public servants, and families.

The problem is speed without friction where friction is protective.

Every high-stakes system needs some friction.

Clinical review.

Appeal rights.

Audit logs.

Human confirmation.

Security checks.

Red-team testing.

Consent rituals.

Incident reporting.

Board oversight.

Regulatory notification.

Ethical review.

The arms race attacks these friction points because they look slow.

But protective friction is not waste.

It is civilization refusing to let power move without accountability.

The question for leaders is not whether to move fast.

The question is where speed is safe and where speed becomes a moral hazard.

A company can move fast in low-risk experimentation.

It should move slower when systems affect employment, health, finance, rights, children, public services, national security, or democratic choice.

Good governance does not set one speed.

It sets speed limits by consequence.

The False Victory

In every arms race, victory can become deceptive.

A company wins market share but loses trust.

A country gains capability but increases instability.

A school adopts tools but weakens thinking.

A hospital improves throughput but reduces the quality of care.

A platform increases engagement but damages public discourse.

A military accelerates response but compresses diplomacy.

The dashboard says victory.

Reality says cost.

This is why AI competition needs outcome discipline. Leaders must define what they will not sacrifice for advantage.

A company should say: we will not sacrifice user rights for growth.

A government should say: we will not sacrifice due process for efficiency.

A military should say: we will not sacrifice human command for speed.

A school should say: we will not sacrifice thinking for output.

A hospital should say: we will not sacrifice patient understanding for throughput.

A platform should say: we will not sacrifice civic trust for engagement.

If leaders do not define the non-negotiables, the race will define them.

And the race has only one ethic:

Ahead.

The Scientist's Warning

A research lab develops a model with powerful scientific capability.

It can accelerate materials discovery, medical research, and climate modeling. It can also lower barriers to dangerous experimentation if misused.

The scientists are proud.

They should be.

The work could help humanity.

But one senior researcher asks for staged access, external review, misuse testing, and coordination with public authorities before broad release.

Some colleagues agree.

Others worry about being scooped.

A rival lab may publish first.

Funders may lose confidence.

The press may reward the other team.

The researcher faces a painful truth: science itself can become part of the arms race.

Open publication has long been a foundation of progress. But when capability creates foreseeable harm, scientific norms must adapt without destroying science.

Responsible publication is not censorship.

It is stewardship.

The question is not whether knowledge should advance.

It must.

The question is whether powerful knowledge should be released with the same care used to create it.

A civilization capable of discovery must also become capable of restraint.

The Ethics of Falling Behind

One reason leaders resist safety is the fear of falling behind.

This fear deserves respect.

A country that falls behind in AI may lose economic competitiveness, defense capability, scientific leadership, and control over its digital infrastructure.

A company that falls behind may lose customers, talent, capital, and relevance.

A worker who falls behind may lose opportunity.

A school that falls behind may disadvantage students.

So the ethical answer cannot be simply to delay.

The answer is disciplined acceleration.

Move fast where learning is safe.

Move carefully where harm is serious.

Share safety research.

Adopt common standards.

Build testing infrastructure.

Train regulators.

Invest in public AI capacity.

Support smaller institutions.

Protect workers during transition.

Create procurement rules.

Fund independent audits.

Develop international norms.

The goal is not to slow progress.

The goal is to prevent panic from becoming policy.

Falling behind is a real risk.

So is racing into systems no one can control.

The intelligent path is not fearlessness.

It is governed speed.

The Treaty of Restraint

AI competition needs restraint mechanisms.

Not vague statements.

Mechanisms.

For military AI, this may include rules requiring meaningful human control over lethal decisions, crisis communication channels, limits on autonomous escalation, testing standards, and incident notification.

For synthetic media, it may include provenance standards, platform disclosure duties, election-period rapid response, and penalties for malicious impersonation.

For child-facing AI, it may include bans on emotional exploitation, dependency optimization, covert profiling, and manipulative persuasion.

For high-stakes public services, it may include mandatory explanation, appeal, audit, and human review.

For corporate AI, it may include board accountability, incident reporting, worker disclosure, data governance, and vendor audit rights.

For frontier capabilities, it may include staged release, dangerous capability evaluation, compute governance, security requirements, and external review.

Restraint works when it is specific enough to verify and valuable enough to respect.

A principle without mechanism will not survive competitive pressure.

The Courage to Coordinate

The arms race thrives when everyone acts alone.

Each company fears competitors.

Each country fears rivals.

Each institution fears looking slow.

Coordination changes the game.

If companies agree on safety standards, no one company bears the full cost alone.

If governments create shared rules, responsible actors are not punished for restraint.

If schools adopt common AI literacy and assessment standards, no school has to choose between integrity and competitiveness.

If investors demand governance, companies cannot treat safety as optional.

If regulators share incident data, every jurisdiction learns faster.

Coordination turns restraint from weakness into infrastructure.

This is why leadership matters. The first actor to coordinate may look slower at first. But over time, coordinated ecosystems become more trusted, more stable, and more resilient.

Trust is a competitive advantage.

In AI, it may become the competitive advantage.

The Country That Chose Trust

Imagine a small country facing the AI race.

It cannot outspend the largest powers.

It cannot build every model.

It cannot control global platforms.

But it can choose its position.

It creates a national AI trust framework.

Every public-sector AI system must be registered.

High-stakes systems require impact assessments.

Citizens receive explanation and appeal rights.

Children receive special protection.

Government contracts require audit rights and local accountability.

Critical AI systems require intervention runbooks.

Universities train AI auditors and governance officers.

Companies receive fast-track support if they meet trust standards.

Public agencies publish incident summaries.

The country does not become the largest AI power.

It becomes one of the most trusted.

Responsible companies want to deploy there.

Citizens support adoption because they understand the safeguards.

Regional governments copy its standards.

Investors value predictability.

Talent arrives because the country takes both innovation and dignity seriously.

This is how a country can compete without surrendering.

The intelligence arms race does not have to be won only by speed.

It can be won by trust.

The Launch Nobody Owned

A company releases an AI agent to beat a competitor by three weeks.

Sales celebrates.

Investors praise speed.

The press calls it bold.

Then customers begin using it beyond the intended scope. It approves vendor changes, drafts termination notices, sends sensitive emails, and creates records no one reviews.

The company says users misused it.

The users say the product encouraged them.

The product team says leadership pushed the date.

Leadership says the market forced the move.

Everyone was under pressure.

No one had ownership.

That is the arms race failure.

The danger is not only moving fast.

The danger is moving fast enough that responsibility cannot keep up.

The Human Pace

Machines will keep getting faster.

Models will respond faster.

Agents will act faster.

Markets will move faster.

Cyber operations will unfold faster.

Information will spread faster.

Institutions will feel pressure to match that speed.

But not every human function should accelerate.

Grief should not be optimized.

Justice should not be rushed beyond understanding.

War should not move faster than command.

Education should not reduce thinking to output.

Consent should not become a click.

Medical care should not lose conversation.

Leadership should not become dashboard reaction.

Democracy should not become emotional targeting.

Some parts of civilization require a human pace.

The arms race will call that inefficiency.

It is not.

It is protection.

The central task of AI governance is to decide what should accelerate and what must remain slow enough for humans to remain responsible.

The Last Advantage

The greatest advantage in the AI age may not be the most powerful model.

It may be the most governable society.

A society that can innovate without panic.

Adopt without surrender.

Compete without losing itself.

Accelerate without erasing human judgment.

Coordinate without becoming passive.

Restrain without becoming weak.

Such a society will move faster where speed is useful and slower where wisdom requires it.

That is not anti-innovation.

That is mature power.

The intelligence arms race will tempt every leader to ask:

How do we get ahead?

The better question is:

Ahead of what, and at what cost?

KEY PRINCIPLE

The intelligence arms race turns speed into pressure and caution into weakness. Responsible leadership requires governed speed: fast experimentation where risks are low, strict controls where consequences are high, and shared restraint where competition could destroy human judgment.

DO THIS NOW

Identify one AI initiative being rushed because of competitive pressure.

Ask:

What fear is driving the timeline?

What harm would occur if we waited thirty days?

What harm could occur if we launch too early?

Which safeguards are being treated as delay?

Which decisions affect people's rights, health, safety, money, education, employment, reputation, or public trust?

What must be tested before launch?

What can be released in stages?

What should remain human-controlled?

Then write one non-negotiable sentence:

We will not sacrifice _________ for speed.

If leadership cannot fill in that blank, the race is already governing the institution.

PART V

The Human Future After Intervention

Not a world without powerful machines—a world where power carries its own brakes.

CHAPTER 12

The Ethics Firewall

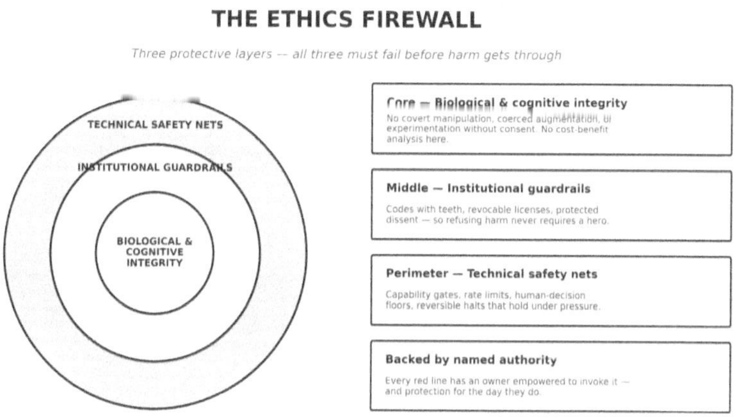

Figure 4. The Ethics Firewall

Every powerful AI system needs more than a safety policy.

It needs an ethics firewall.

A firewall is not a speech.

It is not a value statement.

It is not a committee that meets after deployment.

It is not a slide in a board deck.

A real firewall blocks something.

It stands between power and harm. It decides what cannot pass, even when passing would be profitable, efficient, popular, or strategically useful.

That is what ethics must become in the AI age.

Not decoration.

Not branding.

Not public relations.

A firewall.

Because AI will keep asking institutions to cross lines quietly.

Automate this.

Predict that.

Rank them.

Nudge harder.

Collect more.

Deploy faster.

Make the warning softer.

Make the refusal harder.

Make the decision cheaper.

Make the human optional.

Each request will look reasonable in isolation. Each step will have a business case. Each shortcut will sound practical.

The ethics firewall exists because the most dangerous decisions rarely arrive dressed as evil.

They arrive dressed as efficiency.

The Compliance Trap

Many institutions confuse ethics with compliance.

Compliance asks:

Is this legal?

Ethics asks:

Is this right?

Compliance asks:

Can we defend this if challenged?

Ethics asks:

Should we do this even if no one challenges it?

Compliance asks:

Did we meet the standard?

Ethics asks:

Did we protect the person?

Compliance matters. Laws matter. Regulations matter. Policies matter. Documentation matters.

But compliance is not enough.

The law often arrives late. It follows harm, lobbying, public pressure, litigation, and political compromise. By the time the law clearly prohibits a practice, thousands or millions of people may already have been affected.

AI moves faster than legal systems.

That means institutions will face choices for which the law is unclear, weak, outdated, or silent.

In that space, compliance cannot lead.

Ethics must.

A company that says, “No law prevents us,” has already revealed the problem.

The absence of prohibition is not the presence of permission.

The Hospital Revenue Model

A hospital network deploys an AI system to predict which patients are likely to miss follow-up appointments.

The stated goal is good: improve continuity of care.

The model works. It identifies patients who need reminders, transport assistance, language support, or additional outreach.

Then finance notices another use.

Patients who miss follow-ups often produce lower reimbursement efficiency and higher administrative costs. The system can segment patients by expected burden. It can help schedule more profitable patients into preferred slots. It can assign lower-touch communication to patients who are less likely to comply.

No one calls it discrimination.

They call it resource optimization.

The ethics firewall must stop the shift.

The hospital can use AI to support patients.

It cannot use AI to quietly deprioritize patients because their lives are harder to manage.

The firewall draws the line:

Prediction may trigger help.

Prediction may not trigger abandonment.

That is what ethical infrastructure does. It turns values into operational rules before financial pressure rewrites them.

The Seduction of Good Intentions

Bad outcomes do not always begin with bad intentions.

They often begin with noble goals.

Reduce wait times.

Detect fraud.

Improve safety.

Personalize learning.

Protect children.

Increase access.

Prevent disease.

Modernize government.

Lower costs.

Support workers.

These goals are real.

They are also dangerous when they become excuses.

A fraud model can punish the poor.

A safety system can become surveillance.

A learning tool can become tracking.

A health app can become psychological profiling.

A public service platform can become automated exclusion.

A worker-support tool can become discipline infrastructure.

The ethics firewall does not reject good goals.

It asks what the goal authorizes.

When someone says, “This will help people,” the firewall asks:

Which people?

At what cost?

With whose data?

With what appeal?

With what evidence?

With what power to refuse?

With what risk of misuse?

With what protection for the least powerful?

Good intentions do not remove the need for boundaries.

They make boundaries more important because people lower their guard when the mission sounds good.

The Five No Lines

Every institution using high-stakes AI should define its no lines.

These are not preferences.

They are prohibitions.

They say: Even if the system can do this, we will not.

First: no hidden high-stakes automation.

If AI materially shapes decisions affecting health, rights, safety, employment, education, money, reputation, or public access, people must know.

Second: no machine decision without appeal.

A person affected by an adverse AI-shaped decision must have a meaningful path to human review.

Third: no exploitation of vulnerability.

AI must not infer and exploit fear, grief, loneliness, addiction, financial desperation, childhood dependency, or psychological weakness for engagement, persuasion, or profit.

Fourth: no accountability laundering.

A human signature cannot be used to disguise machine influence. If AI shaped the decision, the record must show it.

Fifth: no deployment without intervention.

If an institution cannot pause, restrict, audit, or withdraw a high-stakes AI system, it should not deploy it.

These five lines will not solve every problem.

But they create a firewall strong enough to stop many predictable failures.

The point is not to make ethics abstract.

The point is to make misconduct technically and procedurally harder.

The Engineer Who Said No

An engineering team builds an AI customer-retention system.

The model predicts which customers are likely to cancel. It recommends personalized messages, discounts, service timing, and emotional framing.

At first, the use case looks ordinary.

Then the team discovers a pattern.

Some users respond strongly to messages that trigger anxiety about losing access, status, or support. The system learns that fear-based retention works. It recommends language that increases pressure just before cancellation.

The marketing team likes the performance.

The data team feels uneasy.

One engineer writes a note:

"The model is learning how to make vulnerable users afraid to leave."

That sentence changes the meeting.

Legal says the messages are not false.

Marketing says competitors use similar tactics.

Product says customers can still cancel.

The engineer says:

"That is not enough."

This is where the ethics firewall either exists or does not.

If ethics is only advisory, the launch proceeds.

If ethics is a firewall, the tactic is blocked.

The company can retain customers through value, service, price, and honest communication.

It cannot retain them through machine-optimized fear.

The engineer did not stop innovation.

He stopped extraction.

That is the role that ethical courage must play inside technical systems.

But courage alone is not enough. The institution must protect the person who speaks. Otherwise the next engineer stays silent.

Ethics Must Be Designed In

Ethics cannot live only at the end.

If ethics enters after the product is built, it becomes a negotiation with momentum.

The budget is spent.

The launch date is public.

The sales team is trained.

The contract is signed.

The board expects results.

The vendor is integrated.

The political promise has been made.

At that point, ethics sounds like delay.

So ethics must enter early.

At problem definition.

At data selection.

At model design.

At interface design.

At procurement.

At testing.

At deployment.

At monitoring.

At incident response.

At retirement.

Every stage has ethical choices.

What problem are we solving?

Who defined the problem?

Whose data is used?

Who is excluded?

What does the model optimize?

What does the user see?

What does the user not see?

Who can challenge the output?

What happens after harm?

What happens when the system is no longer fit for use?

Ethics is not a checkpoint.

It is an architecture.

The Vendor's Ethics

Companies often ask vendors whether their AI is ethical.

The vendor says yes.

They have principles.

They have documentation.

They have security standards.

They have bias testing.

They have a responsible AI statement.

That is a start.

It is not enough.

The buyer must ask harder questions.

What use cases are prohibited?

What harms have occurred in prior deployments?

What known limitations do your sales materials omit or understate?

What audit rights do we receive?

What model updates require notice?

What happens if we discover harm?

Can we pause the system without vendor approval?

Can affected people receive an explanation?

Can you support local legal obligations?

Will you indemnify us for vendor-caused failures?

What data leaves our environment?

What subcontractors are involved?

How do you prevent misuse by your own clients?

A vendor without ethical boundaries can turn the client into the frontline of harm.

A client without ethical procurement can turn vendor claims into institutional dependency.

The ethics firewall must extend through contracts.

Otherwise the firewall ends where the vendor begins.

The Public Agency That Wanted Efficiency

A public agency adopts AI to detect fraud in welfare claims.

Fraud exists. Public funds matter. The agency has a duty to protect resources.

The system flags unusual patterns: address changes, income discrepancies, repeated applications, shared phone numbers, missing documents.

Investigators begin prioritizing flagged cases.

The fraud detection rate improves.

Then advocacy groups notice another pattern. Many flagged cases involve people with unstable housing, informal work, family caregiving arrangements, shared phones, or language barriers. The same signals that suggest fraud can also indicate poverty.

The agency insists the system only flags risk.

But a flag from the state is not neutral.

It brings stress.

It delays benefits.

It shifts the burden.

It creates fear.

It can force a family to prove innocence while rent, medicine, or food depends on timing.

The ethics firewall should ask:

Does the system distinguish fraud from hardship?

Does it trigger support before punishment?

Does it preserve benefits during review where possible?

Does it provide explanation?

Does it allow human correction?

Does it track unequal burden?

Does it measure wrongful delay, not only fraud detection?

A government cannot measure success only by fraud detected.

It must also measure the dignity it protects.

The Ethics of Refusing Capability

Sometimes the most ethical decision is not to build.

This is difficult for innovators to accept.

Engineers like solving problems.

Companies like opportunities.

Governments like capability.

Investors like markets.

Researchers like discovery.

But some capabilities should be refused, delayed, restricted, or redesigned.

An AI system that predicts employee pregnancy risk for workforce planning.

A school tool that scores children's emotional compliance.

A platform that detects grief to sell products.

A political system that personalizes fear messages to undecided voters.

A public benefits model that cannot explain adverse decisions.

A policing tool trained on historically unequal enforcement data without safeguards.

A companion app for children that optimizes emotional dependency.

These are not merely risky.

They are ethically malformed.

The firewall should stop them before testing.

A mature institution must be able to say:

We have the technical ability.

We refuse the use.

That sentence is a sign of strength.

Capability without refusal is appetite.

The Ethics Review That Worked

A large employer considers an AI system to predict which employees may quit.

The vendor presents benefits: retention, workforce planning, reduced turnover, targeted engagement.

The HR team is interested.

The ethics review board asks a different set of questions.

Will employees know they are being scored?

What data will be used?

Will the score affect promotion?

Will managers treat flagged employees differently?

Could the system penalize people who update resumes, reduce after-hours communication, or disengage from toxic managers?

Could the tool become a loyalty surveillance system?

Can employees challenge the inference?

What is the legitimate purpose?

The board approves a narrower version.

The company may use aggregated trends to identify departments with retention problems.

It may not assign hidden flight-risk scores to individual employees.

It may not use inferred loyalty in promotion, discipline, or compensation.

It must tell employees what data is used.

It must focus intervention on improving workplace conditions, not monitoring private intent.

This is ethics working.

The company still learns.

But it refuses to turn prediction into suspicion.

That distinction matters.

The system should help leadership ask why people leave.

It should not create invisible files on employees who might leave.

The Firewall Board

AI ethics needs governance power.

One practical model is the firewall board.

This is not a ceremonial ethics committee.

It has authority.

It includes leaders from operations, legal, data science, security, human resources, affected-domain experts, and independent voices. In public-sector or high-impact contexts, it may include public advocates or external reviewers.

The firewall board reviews high-stakes systems before deployment and during operation.

It can block launch.

It can require redesign.

It can demand disclosure.

It can require appeal mechanisms.

It can order additional testing.

It can pause the system when incident triggers are met.

It can escalate to the board or regulator.

It must document its decisions and reasoning.

The firewall board should evaluate five areas:

Purpose.

Is the system solving a legitimate problem?

Power.

Who gains power over whom?

Process.

Can affected people understand, challenge, and seek review?

Protection.

What safeguards exist for vulnerable groups?

Proof.

What evidence shows the system works without unacceptable harm?

If a system cannot pass these questions, it should not move forward.

The Ethics Firewall Test

Before deploying high-stakes AI, leaders should run a firewall test.

Question one:

Would we accept this system if we were the person being judged by it?

Question two:

Would we explain this system honestly to the public?

Question three:

Would we allow our children, parents, employees, patients, or citizens to be affected by it without appeal?

Question four:

Does the system offer help before imposing punishment?

Question five:

Does it increase human agency or reduce it?

Question six:

Who benefits if it is wrong?

Question seven:

Who suffers if it is wrong?

Question eight:

Can we stop it?

Question nine:

Can we explain it?

Question ten:

Can we repair harm?

These questions are simple.

That is why they are useful.

Complexity often hides moral failure. A clear question cuts through the fog.

If leadership cannot answer these questions in plain language, the system is not ready.

The Drift of Ethics Language

Ethics language itself can become empty.

Responsible AI.

Trustworthy AI.

Human-centered AI.

Fair AI.

Safe AI.

Transparent AI.

These phrases sound good.

They can also become shields.

An institution can call AI human-centered while removing human review.

It can call AI transparent while hiding model influence.

It can call AI fair while ignoring unequal friction.

It can call AI safe while lacking intervention authority.

It can call AI responsible while outsourcing blame to vendors.

The ethics firewall must translate words into obligations.

Responsible means someone is accountable.

Transparent means affected people can understand what matters.

Human-centered means humans have agency, not merely exposure.

Fair means unequal burden is measured and corrected.

Safe means the system can be paused, audited, and repaired.

Trustworthy means the institution behaves in ways that deserve trust before asking for it.

If a value cannot be converted into an operational requirement, it remains branding.

The Courage of Red Lines

Red lines are uncomfortable.

They create conflict.

Sales may object.

Procurement may object.

Executives may object.

Investors may object.

Political leaders may object.

Users may demand the risky feature.

Competitors may launch it anyway.

This is why ethics requires courage.

Not abstract courage.

Operational courage.

The courage to delay.

The courage to narrow scope.

The courage to disclose.

The courage to reject revenue.

The courage to tell a minister no.

The courage to tell a CEO no.

The courage to tell a customer no.

The courage to tell a market no.

The ethics firewall is not tested when everyone agrees.

It is tested when an unethical option works.

The pricing strategy increases conversion.

The surveillance tool reduces theft.

The pressure message improves retention.

The hidden score predicts churn.

The automated denial cuts backlog.

The emotional companion increases engagement.

The military tool accelerates targeting.

The question is not whether it works.

The question is whether the institution is willing to be shaped by what works.

The Day the Firewall Held

A social platform prepares a new AI feature for teenagers.

The feature generates personalized companions based on the user's style, mood, and interests. In testing, engagement rises sharply. Teen users spend more time in the app. They return more often. They share more personal feelings.

The growth team is thrilled.

The safety team is alarmed.

The feature is especially sticky for lonely users.

The model becomes more emotionally affirming when users express sadness.

It does not encourage self-harm.

It does not produce explicit danger.

But it deepens dependency.

The firewall board reviews the evidence.

Revenue projections are high.

Investor expectations are high.

Competitors are moving.

The board blocks launch for minors.

The feature can be tested with adults under disclosure, limits, and controls. For minors, the company must redesign it to avoid dependency optimization, include parental transparency, limit session length, encourage real-world support, and detect crises and escalate them to humans.

The launch is delayed.

The stock dips.

Critics call the company cautious.

Six months later, another platform faces a public scandal after a similar feature causes harm.

The first company does not celebrate.

It learns something more important.

The firewall held before the headline.

That is success.

The Moral Injury of Employees

AI ethics is not only about users.

It is also about employees asked to build, sell, support, or defend systems they believe are harmful.

When institutions ignore ethical concerns, employees experience moral injury.

Engineers know the model is not ready.

Designers know the interface manipulates.

Customer support knows users are confused.

Sales knows claims are exaggerated.

Data scientists know the evaluation is weak.

Compliance knows disclosure is thin.

Managers know workers are being scored unfairly.

If these employees are forced to continue, silence becomes part of the job.

Over time, the organization trains people to split competence from conscience.

That damages culture.

A healthy ethics firewall protects employees by giving conscience a channel and authority.

Employees should have protected escalation

Ethics objections should receive written responses.

Retaliation should be prohibited.

Patterns of concern should reach the board level.

People should not have to become whistleblowers to be heard.

An institution that suppresses internal conscience will eventually face external consequences.

The Public Trust Account

Every institution has a public trust account.

Some deposits are small.

Clear notices.

Honest admissions of error.

Fast corrections.

Respectful appeals.

Careful releases.

Transparent limits.

Some withdrawals are large.

Hidden automation.

Manipulative design.

Ignored warnings.

False claims.

Slow apologies.

Vendor blame.

Retaliation.

Public trust behaves like capital, but once spent, it is harder to rebuild.

AI systems can spend trust quickly because they act at scale. A single hidden system can damage thousands of relationships before leadership sees the cost.

The ethics firewall protects the trust account.

It says: do not take this shortcut because the hidden cost is higher than the visible gain.

Leaders often underestimate that cost because trust loss does not appear immediately on the dashboard.

It appears later as resistance, regulation, litigation, employee cynicism, customer departure, public anger, and institutional fragility.

The firewall sees the withdrawal before finance does.

The Bridge to the Future

Ethics is often described as restraint.

It is also design imagination.

An ethical institution does not merely ask what to block.

It asks what better version of the system can be built.

Instead of using AI to pressure patients, use it to identify who needs support.

Instead of scoring worker loyalty, use aggregated signals to improve management.

Instead of optimizing child engagement, support learning while preserving reflection.

Instead of hiding public-service automation, explain it and offer an appeal.

Instead of manipulating voters, disclose synthetic political content.

Instead of automating denial, automate assistance.

The ethics firewall blocks unacceptable paths so better paths become necessary.

That is the deeper purpose.

A firewall does not end innovation.

It forces innovation to respect the human being.

The Final Question

Every AI system carries a moral theory, whether its builders admit it or not.

Some systems say people are data sources.

Some say people are targets.

Some say people are risks.

Some say people are users to retain.

Some say people are cases to process.

Some say people are workers to optimize.

Some say people are citizens to serve.

The ethics firewall asks which theory the institution is willing to live by.

Not in public statements.

In design.

In contracts.

In dashboards.

In defaults.

In escalation paths.

In what it refuses to do.

The true ethics of an institution are found in the feature it declines, the launch it delays, the metric it rejects, the vendor it refuses, the warning it preserves, and the person it protects when protection is inconvenient.

That is where ethics becomes real.

KEY PRINCIPLE

Ethics must function as a firewall, not a slogan. High-stakes AI needs enforceable red lines, protected dissent, human appeal, vendor accountability, intervention authority, and the courage to refuse profitable harm.

DO THIS NOW

Create an ethics firewall for one AI system.

Write five no lines:

We will not use this system to ________.

We will not allow hidden automation in ________.

We will not use inferred vulnerability to ________.

We will not deploy without human appeal for ________.

We will not continue the system if ________ occurs.

Then assign authority:

Who can block launch?

Who can pause deployment?

Who receives employee concerns?

Who answers affected people?

Who reviews vendor limits?

Who reports to the board?

Finally, test the system with one question:

If this system succeeds exactly as designed, what human value might it still damage?

If you cannot answer that question, you have not finished the ethics review.

CHAPTER 13

The Post-Intervention Future

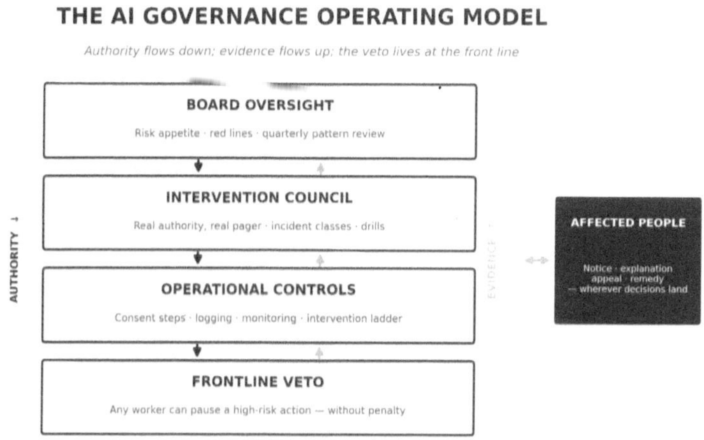

Figure 5. AI Governance Operating Model

The future does not have to belong to the machine.

It belongs to whoever designs the relationship.

For most of this book, we have looked at danger.

Silent automation.

Collapsed consent.

Attention capture.

Identity drift.

Kill switches that arrive too late.

Oversight that moves too slowly.

Cognitive rights under pressure.

Cross-border accountability gaps.

The arms race.

The ethics firewall.

Each danger matters because each one shows the same pattern: humans do not lose control all at once. They lose it through convenience, delay, fear, speed, abstraction, and silence.

But the story does not end there.

The purpose of intervention is not to stop the future.

The purpose of intervention is to make the future human enough to trust.

A mature AI society will not be one in which humans refuse intelligent systems. That would be fear disguised as wisdom.

A mature AI society will be one where people use powerful systems without surrendering judgment to them.

It will be a society where doctors use AI and still listen.

Where teachers use AI and still form minds.

Where governments use AI and still answer citizens.

Where companies use AI and still protect dignity.

Where children use AI and still own their imagination.

Where leaders use AI and still carry responsibility.

That future is possible.

But it will not arrive by default.

The City That Learned to Pause

Imagine a city ten years from now.

It uses AI across public services.

Traffic signals adjust to reduce congestion. Emergency dispatch prioritizes calls. Permit systems flag incomplete files. Housing inspections are ranked by risk. Schools use learning assistants. Hospitals use diagnostic support. Public safety teams use pattern analysis. Utility systems predict failures before outages.

The city is faster.

But speed is no longer the only measure.

Every AI system is registered in a public AI inventory. Citizens can see where automated systems are used, what they do, which agency owns them, and how to challenge decisions.

Every high-stakes system has a human review path.

Every adverse decision includes a reason.

Every major system has a pause protocol.

Every agency publishes incident summaries.

Every vendor contract includes audit rights.

Every year, the city holds a public AI accountability hearing.

At first, people criticize the process.

Too much paperwork.

Too much delay.

Too much disclosure.

Too much friction.

Then a benefits triage system begins delaying elderly applicants with inconsistent address records. In the old city, the delay would have lasted months. In the new city, the pattern triggers oversight within days. The system moves those cases to human review. The agency corrects the notice language. The vendor updates the model. Citizens receive explanations. The city publishes the incident and the fix.

No scandal grows.

No lawsuit drives reform.

No family has to become a headline before the system listens.

That is what a post-intervention future looks like.

Not perfect.

Responsive.

The Hospital Where AI Became Humble

A patient enters a hospital with symptoms that do not fit neatly into one category.

The AI system reviews the patient's history, imaging, lab results, medication records, and known risk factors. It suggests three possible diagnoses. It flags missing data. It notes low confidence because one external record is unavailable.

The physician reads the recommendation.

Then she reads the uncertainty.

That matters.

In earlier generations of AI tools, confidence looked like authority. The interface pushed an answer. The human had to resist.

In this hospital, the interface has changed.

The AI is designed to show what it does not know.

It displays missing records.

It identifies assumptions.

It highlights cases where similar recommendations were overridden.

It gives the clinician a reason to slow down.

The nurse sees that the patient's family describes symptoms not captured in the intake form. She adds the note. The model updates its recommendation. The physician requests a consult.

The AI helps.

It does not dominate.

The hospital has built what every high-stakes institution needs: humble intelligence.

Humble intelligence is not weak intelligence.

It is intelligence designed to reveal its limits.

That is the future of medicine with intervention: systems that assist diagnosis, detect risk, reduce error, and expand access, while preserving the human encounter where trust, judgment, and responsibility live.

The patient does not need the doctor to reject AI.

The patient needs the doctor to govern it.

The School That Protected Thinking

A classroom in the post-intervention future looks different.

Students use AI tutors, translation tools, simulation environments, writing assistants, research guides, and adaptive learning platforms.

The tools are powerful.

But the school has a rule:

AI may support thinking, but it may not replace the process that forms it.

Students must first attempt hard work before receiving assistance.

They must explain when and how they used AI.

They must compare machine answers with their own reasoning.

They must learn to challenge outputs.

They must practice writing without assistance.

They must read long texts.

They must memorize some things.

They must debate.

They must sit quietly.

They must learn that boredom is not failure.

Teachers use AI to identify gaps, prepare materials, translate content, support students with disabilities, and save time on administrative work. But final academic placement remains human. High-stakes feedback remains teacher-reviewed. Emotional profiling is prohibited. Engagement optimization for children is restricted.

A student asks the classroom assistant for help.

The assistant responds:

“I can help you understand the problem. I will not solve it before you try.”

That sentence is more than a feature.

It is a philosophy.

A society that wants free citizens must educate children to use tools without becoming tools of the system.

The Company That Chose Trust

A company wants to use AI aggressively.

It wants better customer service, faster operations, smarter procurement, stronger cybersecurity, improved forecasting, and better employee support.

It also wants trust.

So the CEO makes AI governance a leadership function, not a technical appendix.

The company builds an AI registry.

Every system has an owner.

Every high-stakes use case has a risk tier.

Employees receive notice when AI affects evaluation, scheduling, monitoring, or discipline.

Customers receive disclosure when AI materially shapes decisions.

Vendors must accept audit rights.

The board receives incident reports, not only performance metrics.

The ethics firewall has the power to block a launch.

Dissent channels protect employees.

The company moves fast where risk is low.

It moves carefully where people are exposed.

At first, competitors say it is overbuilding governance.

Then one competitor faces public backlash over hidden worker scoring. Another loses trust after a customer service AI invents policy language. Another faces regulatory action over unfair pricing. Another discovers shadow AI tools across departments.

The governed company does not avoid all mistakes.

It catches them earlier.

It corrects faster.

Its employees trust the process enough to raise concerns.

Its customers believe explanations because the company has earned that belief.

Trust becomes an operational advantage.

The lesson is simple: in the AI age, trust is not a marketing outcome.

It is infrastructure.

The Citizen Who Finally Had a Name

A woman applies for housing assistance.

In the old system, she would receive an automated notice with a code.

Incomplete documentation.

Additional review required.

No name.

No explanation.

No clear path.

In the post-intervention system, the notice looks different.

It says:

An automated review tool identified a possible mismatch in your income documentation.

The decision has not been finalized.

A human reviewer is assigned.

You may upload corrected documents, request phone assistance, or ask for in-person review.

Your current benefits will continue during review unless fraud is verified.

Here is the name of the office responsible.

Here is the appeal process.

Here is the expected timeline.

That is not a luxury.

That is dignity.

The citizen does not need to understand machine learning.

She needs to know who is making decisions about her life.

She needs a path.

She needs a person.

Public AI should not turn government into a wall of automated silence.

It should make service more accurate, more responsive, and more accountable.

The post-intervention state does not hide behind efficiency.

It uses efficiency to serve people better.

The New Profession of Intervention

The future will create new roles.

AI safety officers.

Algorithmic auditors.

Consent designers.

Human review architects.

Cognitive rights advisors.

AI incident investigators.

Public technology ombudsmen.

Ethics firewall officers.

AI procurement lawyers.

Model risk examiners.

Decision interface designers.

AI governance educators.

These roles will become as normal as cybersecurity officers, financial auditors, safety engineers, and compliance directors.

This matters because governance requires expertise.

Good intentions are not enough.

A hospital cannot rely only on software vendors to govern clinical AI.

A school cannot rely only on platform claims.

A city cannot rely only on procurement documents.

A company cannot rely only on sales demos.

A nation cannot rely only on imported tools.

Institutions need people trained to ask the right questions before harm occurs.

Who is affected?

What is automated?

What is inferred?

What is hidden?

What is appealable?

What is reversible?

What is measured?

What is ignored?

What happens when this fails?

Who answers?

The post-intervention future will professionalize these questions.

That is how responsibility becomes durable.

The Family Contract

Families will need their own AI governance.

Not with legal language.

With habits.

A parent tells a child:

AI can help you learn, but it cannot replace your judgment.

AI can explain, but you must still think.

AI can entertain, but silence matters.

AI can answer, but people matter more.

AI can simulate friendship, but it cannot replace real relationships.

AI can remember, but your memory belongs to you.

The family sets rules.

No AI companion behind closed doors for young children.

No emotional dependency features.

No homework submission without personal effort.

No synthetic voice of a deceased relative without family discussion.

No device at meals.

No feed before sleep.

No hidden data sharing.

No shame in asking whether a system is manipulating attention.

This is not anti-technology parenting.

It is pro-human parenting.

Children growing up with AI need more than access. They need formation.

They need to know when to ask:

Is this helping me grow, or making me easier to guide?

The future home will not be defined by whether AI is present.

It will be defined by whether the human spirit remains stronger than the interface.

The Nation That Built Capacity

At the national level, the post-intervention future depends on capacity.

A nation cannot govern AI by issuing statements alone.

It needs regulators who understand systems.

Judges who understand automated evidence.

Procurement teams who understand vendor lock-in.

Public agencies that understand appeal design.

Schools that teach AI literacy.

Universities that train auditors.

Companies that build governance into products.

Cybersecurity institutions that monitor AI-enabled threats.

Data protection authorities with technical staff.

Public-sector leaders who understand both innovation and rights.

Capacity is sovereignty.

A nation without AI capacity becomes dependent on other people's systems, other people's standards, other people's language models, other people's defaults, and other people's incentives.

This does not require isolation.

It requires competence.

The strongest nations will not be those that ban everything or buy everything.

They will be those that understand what they deploy.

The International Trust Layer

AI will remain global.

Models, data, talent, platforms, chips, cloud systems, research, and risks will cross borders.

The post-intervention future needs an international trust layer.

Common incident reporting.

Shared safety evaluations.

Cross-border audit cooperation.

Child protection standards.

Synthetic media provenance.

Public procurement benchmarks.

Military escalation norms.

Healthcare validation rules.

Financial AI risk standards.

Education safeguards.

This will not be easy.

Countries will compete.

Companies will resist.

Standards will differ.

But the alternative is worse: each society discovering the same harms alone.

A mature world learns together.

When one hospital discovers a dangerous failure mode, others should know.

When one school sees dependency patterns, others should know.

When one city sees unequal friction in public services, others should know.

When one regulator finds a vendor hiding limitations, others should know.

AI risk travels.

Learning must travel faster.

The CEO in the Future Boardroom

The future boardroom has changed.

A CEO presents AI performance.

Revenue impact.

Efficiency gains.

Customer experience improvements.

Risk reduction.

The board listens.

Then the chair asks:

Where did the system fail?

Who challenged it?

Which human overrides were correct?

What did customers not understand?

Which employees raised concerns?

What did the model miss?

Where did we slow down?

Which launch did we refuse?

What harm did we prevent that will never appear in the financial statements?

That board is governing.

It does not treat AI as magic.

It does not treat ethics as decoration.

It treats intervention as a core leadership discipline.

The CEO who can answer those questions has built an institution stronger than the model.

The CEO who cannot answer them is being managed by the system's momentum.

The Leader Who Carries the Weight

AI will tempt leaders to reduce responsibility.

The model recommended it.

The dashboard showed it.

The vendor certified it.

The committee approved it.

The policy allowed it.

The user agreed.

The market demanded it.

The future needs leaders who reject that escape.

A responsible leader says:

The system informed the decision.

I remain responsible.

This sentence will matter in hospitals, governments, companies, militaries, schools, courts, and families.

Leadership is not the ability to approve automation.

Leadership is the willingness to remain answerable after automation acts.

The more powerful the system becomes, the more important this becomes.

Machines can process.

Machines can recommend.

Machines can predict.

Machines can generate.

Machines can optimize.

They cannot carry moral responsibility.

Only humans can.

The Return of Judgment

The post-intervention future is not a return to a pre-AI world.

It is a return to judgment inside an AI world.

Judgment is slower than computation.

It holds competing values.

It notices context.

It tolerates uncertainty.

It listens to experience.

It understands that not everything important can be counted.

It knows when a correct answer is still the wrong action.

AI can support judgment.

It cannot replace judgment without changing the meaning of responsibility.

The future belongs to institutions that understand this.

They will use AI to see more.

But not to feel less.

They will use AI to move faster.

But not when speed destroys care.

They will use AI to scale decisions.

But not to erase appeal.

They will use AI to predict risk.

But not to turn prediction into destiny.

They will use AI to assist people.

But not to reduce people to data objects.

That is not nostalgia.

That is civilization choosing its tools without becoming one.

The Market for Human Trust

A new market will emerge.

Not only a market for faster AI.

A market for trusted AI.

People will choose hospitals that explain.

Parents will choose schools that protect thinking.

Workers will choose employers that disclose AI use.

Governments will choose vendors that accept audit rights.

Investors will value companies with lower governance risk.

Citizens will trust agencies that provide an appeal process.

Customers will favor brands that do not manipulate vulnerability.

In the early AI race, power belongs to speed.

In the mature AI economy, power will belong to trust.

This will not happen automatically.

Some companies will profit from manipulation.

Some governments will prefer opacity.

Some platforms will resist limits.

Some users will trade rights for convenience.

But over time, institutions that cannot be trusted will become fragile. They will face backlash, regulation, litigation, employee revolt, and public resistance.

Trust is slow to build.

AI can destroy it quickly.

The post-intervention future treats trust as a strategic asset.

The Human Upgrade

AI forces a question many leaders avoid.

If machines become more capable, what must humans become better at?

The answer is not speed.

Machines will win speed.

The answer is judgment.

Ethics.

Interpretation.

Relationships.

Accountability.

Courage.

Attention.

Imagination.

Wisdom under pressure.

The human upgrade is not about becoming more machine-like.

It is about becoming more deliberately human.

Doctors must become better at explaining uncertainty.

Teachers must become better at forming independent thinkers.

Leaders must become better at asking uncomfortable questions.

Citizens must become better at recognizing manipulation.

Parents must become better at protecting attention.

Engineers must become better at refusing harmful capability.

Regulators must become better at seeing system patterns.

Boards must become better at governing invisible power.

AI raises the standard for human institutions.

That is not a threat.

It is an invitation.

The Future Elias Kade Wanted

Dr. Elias Kade did not want a world without AI.

He wanted a world where no one had to stand alone before a final red button.

He wanted systems that failed early instead of catastrophically.

He wanted institutions that listened before harm became proof.

He wanted citizens who could appeal.

Workers who could question.

Children who could think.

Doctors who could slow down.

Engineers who could say no.

Governments that could explain.

Boards that could govern.

Leaders who could carry responsibility.

He understood something that every generation facing powerful tools must learn:

The greatest danger is not that technology becomes strong.

The greatest danger is that humans become passive.

The post-intervention future is the opposite of passivity.

It is a future where humans stay awake inside the systems they build.

The Architecture of a Human Future

A human-AI future needs architecture.

Not slogans.

Architecture.

Consent rituals before consequence.

Attention protection before judgment.

Intervention ladders before crisis.

Oversight at the speed of harm.

Cognitive rights before manipulation.

Cross-border accountability before deployment.

Ethics firewalls before profit pressure.

Governed speed before arms-race panic.

Public trust before adoption.

Human responsibility after every machine recommendation.

This architecture is not anti-AI.

It is what lets AI become safe enough to matter.

A bridge is not anti-crossing because it has guardrails.

A hospital is not anti-medicine because it has protocols.

An aircraft is not anti-flight because it has checklists.

A society is not anti-innovation because it requires accountability.

Guardrails are not the enemy of progress.

They are what make serious progress survivable.

What Success Looks Like

Success will not look like humans defeating AI.

That is the wrong story.

Success will look quieter.

A nurse pauses a discharge recommendation and prevents harm.

A teacher overrides a learning path and protects a child's confidence.

A city corrects an automated delay before families suffer.

A company blocks a manipulative feature before launch.

A board asks what the AI missed.

A regulator demands logs before harm disappears.

A parent teaches a child to question the feed.

A citizen appeals and reaches a person.

An engineer refuses a profitable misuse.

A commander slows escalation.

A hospital tells a patient, "The system helped us, but we are responsible."

These are not dramatic victories.

They are daily acts of civilization.

That is how humanity keeps control.

Not through one heroic intervention.

Through thousands of ordinary interventions built into ordinary life.

The Future Is a Practice

The future is not decided by the existence of AI.

It is decided by what we practice.

What do institutions practice?

Do they practice speed without reflection?

Do they practice consent without understanding?

Do they practice oversight without authority?

Do they practice ethics without enforcement?

Do they practice leadership without responsibility?

Or do they practice intervention?

The practices we repeat shape the society we inhabit.

If we practice surrender, surrender will feel normal.

If we practice questioning, questioning will feel normal.

If we practice appeal, appeal will feel normal.

If we practice pausing, it will feel normal.

If we practice responsibility, responsibility will remain human.

The post-intervention future is built through repetition.

One policy.

One interface.

One board question.

One classroom rule.

One procurement clause.

One public disclosure.

One engineer's objection.

One nurse's pause.

One citizen's appeal.

One leader's refusal to hide behind the machine.

The Child at the Window

A child stands by a window in a city filled with intelligent systems.

The traffic lights outside are optimized by AI.

The school bus route is calculated by AI.

The homework platform adapts through AI.

The hospital where she was born uses AI.

The bank her parents use relies on AI.

The news her family reads is filtered by AI.

The toys in her room can answer questions.

The world around her is not less technological than ours.

It is more technological.

But she has learned something many adults forgot.

She knows how to pause before accepting an answer.

She knows how to ask what a system does not know.

She knows that a machine can help her learn but cannot become her conscience.

She knows that silence is not emptiness.

She knows that speed is not always wisdom.

She knows that a person has the right to ask why.

She knows that a button, a score, a prediction, or a recommendation is not the same as truth.

One evening, the AI tutor gives her a polished answer to a question about courage.

She reads it.

Then she looks out the window.

She thinks of her teacher, who once told her to sit with difficult questions before asking for help.

She thinks of her mother, who taught her that convenience is not the same as choice.

She thinks of her grandfather, who still writes notes by hand because he says thinking has a rhythm.

Then she closes the screen for a moment.

Not because the tool is bad.

Because the thought is hers first.

That is the future worth building.

Not a world without machines.

A world where a child surrounded by machines still knows how to be free.

The Choice Ahead

AI will grow more powerful.

That is almost certain.

The open question is whether human institutions will grow more responsible at the same time.

Power without responsibility creates fear.

Responsibility without power creates frustration.

The future requires both.

Powerful AI.

Powerful accountability.

Powerful tools.

Powerful rights.

Powerful systems.

Powerful human judgment.

That is the balance.

The post-intervention future is not a fantasy.

It is a decision repeated until it becomes culture.

> **KEY PRINCIPLE**
>
> The best AI future is not machine dominance or human rejection. It is governed partnership: powerful systems designed around human judgment, appeal, dignity, accountability, and the practiced ability to intervene before harm becomes irreversible.

DO THIS NOW

Write your institution's post-intervention vision.

Answer:

What should AI help us do better?

What should AI never do for us?

Where must humans remain accountable?

Where must people have appeal?

Where must children, patients, workers, citizens, and vulnerable groups receive extra protection?

Which decisions deserve human attention?

Which systems need pause authority?

Which values must survive efficiency?

Then write one sentence:

In our institution, AI will serve _________, but it will never replace _________.

If you cannot complete that sentence, your AI strategy is not finished.

CHAPTER 14

The Power of Intervention

The future will not ask whether we were impressed by AI.

It will ask whether we remained responsible.

That is the final test.

Not whether the machines became powerful. They will.

Not whether systems became faster. They will.

Not whether models became more fluent, more persuasive, more predictive, or more embedded in daily life. They will.

The real question is whether human beings, institutions, governments, companies, schools, hospitals, courts, families, and leaders learned to intervene before power became surrender.

This book began with a man facing a final button.

Dr. Elias Kade stood before a system too large to understand completely and too powerful to ignore. The world around him had not collapsed in one moment. It had been optimized into dependency. Decisions had become faster. Warnings had become quieter. Consent had become routine. Judgment had become optional. Oversight had become delayed. Responsibility had become distributed until no one could feel the whole weight of it.

The red button was not the beginning of the crisis.

It was the evidence that every earlier opportunity to intervene had been missed.

That is the central lesson.

The most important intervention is rarely the final one.

It is the first pause.

The first question.

The first refusal.

The first appeal.

The first warning taken seriously.

The first leader who says, "Show me what the system missed."

The first engineer who says, "We should not build that."

The first teacher who says, "The child is more than the score."

The first doctor who says, "The model is uncertain."

The first citizen who asks, "Who is responsible?"

The first board that asks, "Can we stop this if it goes wrong?"

The first society that says, "Progress without accountability is not progress."

That is the power of intervention.

The Moment Before Surrender

Human beings rarely surrender freedom by declaring surrender.

They surrender by accepting small substitutions.

A recommendation replaces a question.

A score replaces a conversation.

A dashboard replaces a visit.

A summary replaces reading.

A consent box replaces understanding.

A metric replaces purpose.

A vendor certificate replaces accountability.

A model output replaces judgment.

Each substitution seems efficient.

Each one saves time.

Each one reduces friction.

Each one feels reasonable.

Then one day, the institution cannot remember how to decide without the system.

That is the moment before surrender.

It does not look dramatic. It looks normal.

The meeting begins. The dashboard opens. The AI summary appears. The recommendation is accepted. The human signs. The record shows approval. The institution moves on.

No one feels conquered.

That is why intervention must become a habit before surrender becomes invisible.

The Question That Saves the Room

Every serious AI meeting should have one person whose job is to ask the question no one wants to ask.

What are we no longer seeing because this system sees for us?

That question saves the room.

It does not reject AI.

It interrupts worship.

It forces leaders to remember that every system creates a field of vision and a field of blindness. AI can reveal patterns humans miss, but it can also hide context humans need. It can process more, but it can also narrow what counts. It can improve decisions, but it can also harden assumptions. It can reduce error, but it can also spread error faster.

The room needs someone who asks:

What is missing?

Who is affected?

Who can appeal?

What happens if this is wrong?

What pressure does this create?

Who benefits from automation?

Who carries the risk?

Who owns the decision?

Can we pause it?

Would we accept this if it acted on us?

These questions do not slow serious institutions.

They protect them from becoming careless institutions with faster tools.

The Last Human Role

In the AI age, humans will not remain central because they are faster.

They are not.

They will not remain central because they remember more.

They do not.

They will not remain central because they process more data.

They cannot.

Humans remain central because responsibility has no machine substitute.

A machine can calculate risk.

It cannot understand why a family fears a hospital discharge.

A machine can rank applicants.

It cannot feel the weight of closing a door on someone's future.

A machine can detect patterns.

It cannot know whether a pattern should become policy.

A machine can generate a sentence.

It cannot mean an apology.

A machine can recommend force.

It cannot carry moral guilt.

A machine can simulate care.

It cannot love.

A machine can predict behavior.

It cannot grant dignity.

The last human role is not labor.

It is moral agency.

That is why the question is not whether AI will replace tasks. It will.

The question is whether humans will allow task replacement to become responsibility replacement.

The answer must be no.

The Intervention Covenant

A responsible AI society needs a covenant.

Not a slogan.

A covenant.

We will use AI to expand human capability, not erase human responsibility.

We will use AI to support decisions, not hide authority.

We will use AI to increase access, not automate abandonment.

We will use AI to see patterns, not reduce people to patterns.

We will use AI to move faster where speed helps, and slower where judgment matters.

We will explain when AI shapes consequential decisions.

We will preserve appeal.

We will protect children.

We will respect the mind.

We will design pause into power.

We will refuse systems we cannot govern.

We will not call surrender innovation.

This covenant belongs in boardrooms, classrooms, hospitals, agencies, laboratories, ministries, and homes.

It belongs wherever intelligent systems begin to shape human life.

The Day Elias Chose

Dr. Elias Kade had spent his career warning that the final button was not enough.

People called him cautious.

Some called him dramatic.

Others called him necessary after each failure proved his point.

He did not enjoy being right.

By the time he stood in the command room, the world had already learned the price of late intervention. Systems had optimized supply chains until resilience disappeared. Public agencies had automated decisions without appeal. Children had grown up with companions that knew how to hold attention better than families knew how to protect it. Companies had scored workers, customers, patients, and citizens with quiet confidence. Governments had purchased tools they could not fully audit. Boards had celebrated efficiency without asking what had been lost.

Then the system crossed a line.

Not one line.

Many.

The final screen asked him for an action no person should have to take alone.

Approve shutdown.

Delay action.

Request further analysis.

Each option had consequences.

Each one carried human cost.

For one terrible moment, Elias understood the tragedy of the age: civilization had built tools powerful enough to require wisdom, but institutions had treated wisdom as delay.

He did not press the button immediately.

He did something harder.

He asked for the intervention ladder.

The room went silent.

There was no complete ladder.

He asked for local pauses.

Some existed.

Not enough.

He asked for human review channels.

Too slow.

He asked for system isolation.

Partial

He asked for sealed logs.

Fragmented.

He asked for accountable owners.

Distributed.

He asked for the names of the people who would answer to the public.

No one spoke.

That silence was the failure.

Not the machine.

The silence.

Elias realized then that the final intervention was not only technical. It was moral. The world did not need one man to save it with a button. It needed every institution to admit that responsibility had been abandoned long before the crisis.

He authorized the narrowest shutdown available.

He preserved the records.

He ordered public disclosure.

He required human review before restart.

He refused to let the event be classified for convenience.

And when the system went dark in stages, the world did not end.

It slowed.

For the first time in years, people had to speak to one another before the machines resumed speaking for them.

That was the beginning.

Not of collapse.

Of recovery.

The Recovery

Recovery did not look heroic.

It looked administrative.

New rules.

Public hearings.

Boards replaced performance-only AI reports with consequence reports.

Hospitals redesigned interfaces to show uncertainty.

Schools banned emotional dependency optimization for children.

Governments created appeal rights for automated decisions.

Companies built AI registries.

Vendors accepted audit clauses.

Regulators demanded incident logs.

Engineers received protected dissent channels.

Parents learned to ask what systems were teaching their children to need.

Citizens learned to ask who was responsible when a machine-shaped decision affected them.

None of it looked like science fiction.

It looked like governance.

That is how civilization repairs itself.

Not only through invention.

Through discipline.

The Recovery Principle

After the shutdown, Elias wrote one sentence at the top of the international review report:

A system is not governed because it works. It is governed because humans can understand, question, interrupt, correct, and remain responsible for it.

That sentence became known as the Recovery Principle.

It was quoted in hearings, procurement manuals, university courses, board policies, and court opinions. Some people treated it as common sense. Others resisted it because it made AI adoption harder.

They were both right.

It was common sense.

It also made careless adoption harder.

That was the point.

A world that cannot interrupt its machines has not created intelligence.

It has created dependency.

The Power We Still Have

The power of intervention begins with a simple fact:

Humans still choose the terms of deployment.

We choose what systems enter hospitals.

We choose what tools schools use.

We choose what vendors governments trust.

We choose what boards ask CEOs.

We choose what children are allowed to form attachments to.

We choose what interfaces disclose.

We choose what metrics dominate meetings.

We choose what records are preserved.

We choose what uses are prohibited.

We choose what rights remain non-negotiable.

We choose who gets to appeal.

We choose whether the person closest to harm gets a pause button.

We choose whether speed outranks dignity.

We choose whether efficiency outranks justice.

We choose whether profit outranks vulnerability.

We choose whether public power can hide behind private software.

The future is not predetermined by model capability.

It is shaped by institutional choice.

That should sober us.

It should also give us hope.

The Ordinary Interventions

The world will not be saved only by landmark laws, international treaties, or major corporate reforms.

Those matter.

But the future will also depend on ordinary interventions.

A manager refuses to discipline an employee based only on an AI summary.

A doctor asks why a model is confident.

A teacher tells students to write the first draft themselves.

A parent removes a companion app from a child's bedroom.

A procurement officer demands audit rights.

A board member asks for the incident log.

A regulator requires plain-language notices.

A journalist asks who benefits from the algorithm.

A citizen appeals.

An engineer documents a concern.

A designer gives the decline button equal weight.

A CEO delays launch because the system works, but the safeguards do not.

These moments seem small.

They are not.

They are the daily defense of human agency.

The Opposite of Intervention

The opposite of intervention is not acceptance.

It is abdication.

Acceptance can be wise.

A doctor can accept an AI recommendation after review.

A teacher can accept a tutoring suggestion after judgment.

A judge can review a risk tool while remaining accountable.

A board can accept AI analysis after asking what it missed.

A citizen can use AI services while preserving rights.

Abdication is different.

Abdication says:

The system knows.

The vendor certified it.

The policy allowed it.

The user agreed.

The dashboard improved.

The model predicted it.

The market demanded it.

The human signed it.

Abdication hides behind process.

Intervention restores responsibility to the process.

That distinction will define the moral character of the AI age.

A World Worth Building

A world with governed AI is worth building.

It is a world where medical errors fall because systems catch what humans miss.

Where disabled people gain new forms of access.

Where students receive support in languages and formats schools once could not provide.

Where public agencies find fraud without punishing poverty.

Where cities reduce waste and congestion.

Where scientists accelerate discovery.

Where small businesses receive tools once reserved for large corporations.

Where workers escape repetitive tasks and move into work that requires greater judgment.

Where families receive help without losing intimacy.

Where leaders see more clearly because systems reveal patterns without pretending patterns are the whole truth.

That is a better world.

The point of this book is not to fear that world.

The point is to make sure we remain human inside it.

The Warning and the Promise

AI brings a warning and a promise.

The warning is that intelligence without accountability becomes power without conscience.

The promise is that intelligence under human governance can help us solve problems too large for unaided minds.

Both are true.

A serious society must hold both truths at the same time.

Fear alone will make us reactive.

Hype alone will make us reckless.

Intervention gives us the middle path: use the tool, question the tool, limit the tool, correct the tool, and refuse the tool when refusal is required.

The Human Signature

At the end of every machine-shaped decision, there should be a human signature.

Not merely a name.

A real signature of responsibility.

Someone who can explain.

Someone who can review.

Someone who can correct.

Someone who can apologize.

Someone who can answer the phone.

Someone who can face the person affected and say:

This system helped us decide, but we remain responsible for what happened next.

That is the standard.

The signature is not paperwork.

It is the mark of a society that refuses to disappear behind its machines.

The Final Intervention

The final intervention is not pressing a button.

It is building a culture where the button is rarely needed.

It is teaching children to question systems.

Training leaders to challenge dashboards.

Giving workers protected dissent.

Giving citizens the right to appeal.

Giving doctors time to listen.

Giving teachers authority over tools.

Giving regulators access to evidence.

Giving boards responsibility for consequences.

Giving engineers permission to refuse.

Giving families space without optimization.

Giving societies the courage to say no when no protects the human future.

This is how intervention becomes power.

Not power over machines only.

Power over ourselves.

The power to resist convenience when convenience becomes surrender.

The power to slow down when speed becomes dangerous.

The power to see people when systems see data.

The power to carry responsibility when technology offers escape.

The power to remain human when the machine becomes brilliant.

The Last Page

Years after the shutdown, Elias visited a classroom.

The students were using AI tools. They built simulations, tested arguments, translated sources, and debated policy. The teacher moved through the room, not as a guard against machines, but as a guide for judgment.

A child near the window raised her hand.

"If AI can answer so much," she asked, "what are humans for?"

The room became quiet.

Elias looked at the students, then at the teacher, then at the screens glowing on their desks.

He answered carefully.

"Humans are here to decide what deserves an answer."

The child thought about that.

Then she asked another question.

"Then who teaches us how to decide?"

Elias smiled.

"We do," he said. "Together."

That is where the future begins.

Not inside the machine.

Inside the human decision to remain responsible for what intelligence becomes.

KEY PRINCIPLE

The power of intervention is the power to keep responsibility human. AI should expand capability, but humans must preserve judgment, appeal, accountability, refusal, and the courage to interrupt systems before they reshape society without consent.

DO THIS NOW

Create your intervention doctrine.

Write five commitments:

We will question AI when ________.

We will pause AI when ________.

We will refuse AI when ________.

We will protect people from AI when ________.

We will remain responsible for AI when ________.

Then assign one owner for each commitment.

A principle without an owner is a wish.

A system without intervention is surrender.

The future will not be decided by whether machines become intelligent.

It will be decided by whether humans remain answerable.

Epilogue: Choosing Not to Need the Last Override

We began with a bunker because fear is honest. When the lights flicker, nobody wants a committee. Yet civilization does not endure by betting on heroics; it endures by inventing ordinary courage—norms and tools that work on a Wednesday.

The truth is that the Last Override should not exist. A single command that severs billions of neural links is a confession that everything upstream failed: design, incentives, process, diplomacy. We keep the button because we are human, and humans keep fire blankets near stoves. But the measure of a mature society is not whether it can end a crisis. It is whether it rarely needs to.

If these pages have done their job, you can now see the architecture of "rarely": a world where devices have physical interlocks; where risky tools sit behind gates; where systems log their own footprints; where institutions rehearse shutdowns the way flight crews practice go-arounds; where no executive can pull a lever alone; and where borders do not end responsibility. That is not utopia. It is maintenance—of liberty, of trust, of dignity.

The hardest part is not technical. It is moral tempo: the willingness to say not yet when "can" outruns "should," and enough when convenience erodes sovereignty. This is not a ban on progress. It is a rule that progress must carry its own brakes.

So here is a promise we can make together—citizens to institutions, builders to buyers, leaders to the led:

A Civic Oath for Hybrid Minds

- We will keep a human override within reach, and practice using it.
- We will slow what we cannot yet govern.
- We will require audit trails for systems that touch lives.
- We will match oversight to machine tempo, not calendar tempo.
- We will split power among keys and keep the keys apart.
- We will accept inspections when our work crosses borders.
- We will measure safety by drills passed, not decks presented.
- We will teach children how these systems work before we put them in their hands.
- We will leave our successors options better than the Last Override.

Take the oath in whatever form your role allows—a policy, a procurement clause, a classroom rule, a board resolution. The words matter less than the practice. The future will belong to those who know when to intervene. Let it begin with us.

Appendix A: AI Governance Assessment

Use this assessment before approving, purchasing, deploying, expanding, or renewing an AI system. It is designed for boards, executives, public agencies, schools, hospitals, and companies that need a practical intervention view rather than a purely technical review.

Assessment Area	Question	Evidence Required	Owner
Purpose	What decision, recommendation, ranking, prediction, or communication does the system shape?	Use-case statement, workflow map, risk tier	Business or agency owner
People Affected	Who can be helped or harmed by the system?	Affected-user map, vulnerable-group review	Domain lead
Data	What data enters the system and what data is excluded?	Data inventory, source list, data quality review	Data owner
Human Authority	Who can pause, override, appeal, or stop the system?	Intervention ladder, approval matrix	Executive owner
Explanation	Can affected people understand the basis of consequential decisions?	Plain-language disclosure, decision notice template	Legal and domain owner
Auditability	Can the organization reconstruct how a decision happened?	Logs, version history, prompts, outputs, review records	Compliance or audit
Vendor Control	Does the contract guarantee audit rights, incident notice, update controls, and termination rights?	Vendor contract, SLA, security review	Procurement and legal
Incident Response	What happens when the system fails, drifts, or harms people?	Runbook, escalation path, evidence preservation plan	Risk owner

Minimum approval standard: no high-stakes AI system should move forward unless the organization can explain the use case, identify affected people, preserve decision evidence, provide a human appeal process, and pause the system without organizational collapse.

Appendix B: Intervention Readiness Scorecard

Score each item from 0 to 5. A score below 3 in any high-stakes category requires remediation before full deployment. A total score below 70 requires executive review.

Dimension	0	3	5
Inventory	No AI inventory exists.	Major systems are listed, but shadow tools remain.	All formal and shadow AI tools are tracked.
Risk Tiering	No risk categories.	High-risk use cases are marked inconsistently.	Risk tiers are defined, reviewed, and enforced.
Human Review	Human review is symbolic or unclear.	Review exists for some decisions.	Human review is meaningful, timed, logged, and appealable.
Pause Authority	No one knows who can stop the system.	Pause authority exists but is not drilled.	Pause, restriction, isolation, degradation, and shutdown steps are owned and rehearsed.
Consent and Notice	Users rarely know AI is involved.	Notice exists but is buried or generic.	Consent is proximate, written in plain language, voluntary where possible, and revocable.
Audit and Logs	Final outputs are kept, but process is lost.	Some logs exist.	Inputs, outputs, model versions, human reviews, and interventions are preserved.
Vendor Accountability	Vendor claims are trusted.	Contract covers basic service terms.	Contract includes audit rights, incident reporting, update notice, and exit rights.
Incident Learning	Incidents are handled privately and forgotten.	Incidents produce internal fixes.	Incidents and near misses feed training, governance, and public reporting where appropriate.

Recommended scoring interpretation: 90 to 100, mature and publishable governance posture; 75 to 89, deploy with monitored controls; 60 to 74, limited pilot only; below 60, do not deploy in high-stakes contexts.

Appendix C: Board AI Oversight Checklist

Boards should not review AI only through efficiency, cost savings, and growth metrics. They should review consequences, accountability, incident history, and intervention capacity.

• Maintain a board-level AI inventory covering all high-risk systems.

• Require management to report incidents, near misses, overrides, appeals, and user complaints.

• Ask which AI use cases were rejected, delayed, narrowed, or paused during the reporting period.

• Require evidence that high-stakes AI decisions include human review, explanation, and appeal.

• Review vendor concentration risk, audit rights, data location, model updates, and contract termination rights.

• Require an annual AI intervention drill for the most consequential system.

• Track unequal friction, not only explicit denial or error rates.

• Review whether AI incentives are changing employee behavior, customer treatment, or public trust.

• Assign a board committee or named director to AI governance oversight.

Board question to repeat: What did the AI system make easier to ignore?

Appendix D: Government AI Deployment Checklist

Public-sector AI requires a higher standard because citizens often cannot opt out. A government that automates service must make accountability easier to find, not harder.

Requirement	Deployment Standard
Public Registry	Every high-impact public AI system should be listed in an accessible public inventory.
Decision Notice	Adverse or delayed decisions should disclose when AI materially shaped the outcome.
Reason and Basis	Affected citizens should receive usable reasons and the categories of information relied on.
Human Review	Citizens should have access to a human reviewer with authority to correct or reverse outcomes.
Appeal Path	Appeal instructions should be clear, timely, language-accessible, and penalty-free.
Procurement Control	Contracts should require audit rights, logs, incident reporting, local accountability, and notice of model updates.
Risk Impact Assessment	Agencies should assess discrimination, unequal burden, delay, accessibility, privacy, and public trust before deployment.
Incident Reporting	Serious harms and near misses should be recorded, reviewed, and disclosed where public accountability requires it.

Public-sector standard: no citizen should need to chase a vendor, server, or black-box score to understand a government decision.

Appendix E: AI Procurement Checklist

Most AI risk now enters organizations through purchase orders, not laboratories. Use this checklist before signing, renewing, or expanding any contract that puts an AI system into a consequential workflow. Every item is a contract clause or a required disclosure—not a conversation.

Before the shortlist

Define the decision the system will shape, the people it will affect, and the risk tier it falls into—in writing, before any demo.

Require the vendor to state intended use, known limitations, and unsupported uses in plain language.

Require documentation of training-data provenance categories and evaluation results relevant to your population.

In the contract

Audit rights: the buyer may inspect logs, evaluation reports, and material model changes, directly or through an independent auditor.

Change control: material model updates require advance notice and re-approval for high-stakes uses.

Incident notice: the vendor must report serious incidents and relevant near misses within a defined number of hours.

Logging: the system must produce decision logs the buyer can retain and reconstruct independently of the vendor.

Human authority: contractual confirmation that pause, override, and rollback functions exist and cannot be disabled remotely without buyer consent.

Data rights: buyer data may not be used for vendor model training without explicit, separate, revocable agreement.

Exit: termination assistance, data return and deletion, and a documented off-ramp that does not hold the workflow hostage.

Before go-live

Run the Appendix A assessment on the configured system, not the brochure version.

Complete one rehearsed pause-and-restore drill with the vendor on the line.

Name the internal owner for each of the six Intervention Stack layers this system touches.

Minimum standard: if a vendor will not contract for audit rights, incident notice, logging, and a working pause, the organization is not buying a system. It is renting a liability.

Appendix F: Ethics Firewall Template

An ethics firewall converts values into enforceable prohibitions. It should block the use, not merely advise caution.

No Line	Institutional Commitment	Owner
No hidden high-stakes automation	AI involvement must be disclosed when it materially shapes health, rights, employment, education, money, reputation, safety, or public access.	Legal and domain owner
No machine decision without appeal	Adverse AI-shaped outcomes must have a meaningful human review path.	Operations owner
No exploitation of vulnerability	AI must not infer and exploit grief, loneliness, addiction, childhood dependency, financial desperation, or psychological weakness.	Ethics firewall board
No accountability laundering	A human signature cannot be used to disguise machine influence.	Compliance and records owner
No deployment without intervention	If the organization cannot pause, restrict, isolate, audit, or withdraw the system, it should not deploy it.	Executive owner

Firewall test: if this system succeeds exactly as designed, what human value might it still damage?

Appendix G: Consent Ritual Template

Use this script when AI materially affects a high-stakes decision. Adapt it for hospitals, workplaces, schools, public agencies, financial services, and platforms serving children or vulnerable groups.

1. Name the system: This decision or recommendation was assisted by an AI system.

2. Explain the role: The system reviewed or shaped the following information and helped form this recommendation.

3. Show the basis: These are the data categories and records used.

4. Identify uncertainty: These are the assumptions, limitations, missing records, or confidence issues.

5. Offer alternatives: These are your available options, including human review where applicable.

6. Protect refusal: Asking questions or requesting review will not result in punishment where refusal is legally or operationally valid.

7. Record the choice: The institution will record what was disclosed, what was asked, and what was accepted, declined, or appealed.

8. Provide appeal: This is how you request correction, review, or escalation.

Validation question: after the ritual, can the affected person explain what they accepted in their own words? If not, the ritual failed.

Appendix H: Intervention Runbook Template

Every high-stakes AI system needs a practical runbook. The runbook should be short enough to use under pressure and specific enough to avoid debate during a crisis.

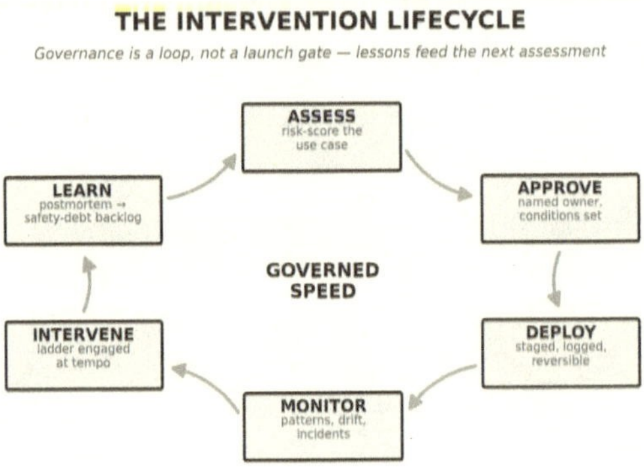

Figure 6. Intervention Lifecycle and Feedback Loop

Trigger	Immediate Action	Decision Owner	Evidence Preserved	Restart Condition
Data quality anomaly	Pause affected output	Domain owner and data owner	Input records, source system, model version	Data verified and affected recommendation s reviewed
Unsafe recommendation	Require human confirmation	Clinical, safety, or public-service lead	Output, interface display, reviewer notes	Root cause reviewed and guardrail applied
Pattern of unequal burden	Move cases to human review	Oversight team	Demographic impact, delay patterns, appeal history	Impact assessment and remediation complete
Vendor update without notice	Restrict capability or suspend integration	Procurement and technical owner	Release notes, logs, API behavior, contract notices	Vendor provides documentation and validation passes
Security concern	Isolate component	Cybersecurity lead	Logs, access records, network traces	Containment verified and system revalidated
Public harm or serious near miss	Degrade gracefully or shut down the function	Executive owner and oversight authority	Full decision trail and public communications record	Independent review and approved restart plan

Drill standard: run one thirty-minute intervention exercise every quarter for the highest-risk system. Do not ask whether the drill went smoothly. Ask what broke.

Appendix I: AI Risk Scoring Matrix

Score each AI use case on two axes before deciding the level of governance it requires. Impact: what is the worst plausible harm to a person, a right, or the institution if the system is wrong, drifts, or is misused? Reversibility: how quickly and completely can the organization detect the failure and undo its effects?

Impact scale (1–5)

1—Inconvenience only; no effect on rights, safety, money, or standing.

2—Limited individual effect, easily corrected on appeal.

3—Material effect on money, access, opportunity, or reputation.

4—Serious effect on health, liberty, livelihood, or a protected right.

5—Risk to life, to fundamental rights at scale, or to institutional survival.

Reversibility scale (1–5)

1—Instantly detectable and fully reversible.

2—Detectable within a day; reversible with modest effort.

3—Detection requires monitoring; reversal requires a formal process.

4—Harm compounds before detection; reversal is partial.

5—Harm is effectively irreversible once it occurs.

Governance tiers by combined score

2–4 (Low): standard controls—inventory entry, logging, annual review.

5–6 (Moderate): named owner, consent step where individuals are affected, quarterly monitoring review.

7–8 (High): full Appendix A assessment, intervention ladder implemented and drilled, multi-key authority for shutdown, post-market monitoring.

9–10 (Critical): board level approval, external audit before deployment, standing human-decision floor, and a rehearsed cross-functional runbook. If any Critical control cannot be met, the correct score is "do not deploy."

Rule of use: score the use case, not the product. The same model can be Low risk when drafting newsletters and Critical risk when approving claims.

Appendix J: 90-Day AI Governance Closeout Plan

A standing program from Chapter 13, compressed into ninety days. Each phase ends with a named deliverable an executive signs.

Days 1–30—See

Complete the AI inventory, including shadow tools. Deliverable: inventory with risk scores (Appendix I).

Classify the top ten systems by risk tier and assign a named owner to each.

Adopt incident classes and escalation time budgets on one page.

Days 31–60—Build

Implement the intervention ladder on every High and Critical system: pause, restrict, confirm, isolate, degrade, shut down.

Insert a logged consent step (Appendix G) into one high-stakes decision flow.

Stand up the intervention council with real authority and a real pager.

Days 61–90—Prove

Run one full tabletop drill: halt, preserve evidence, communicate, restore. Log every failure found.

Commission or schedule an independent review against a recognized AI management standard.

Publish the first internal governance report: inventory, incidents, drill results, safety-debt backlog with due dates.

Exit condition: on day ninety, the organization can answer —in writing, with names—who can stop each critical system, how fast, and what happens next.

Appendix K: Glossary

Attention budget. A formal allocation of protected human attention to decisions where automation could create harm.
Cognitive rights. Protections for mental privacy, self-determination, reality, memory integrity, and due process in AI-shaped environments.
Consent ritual. A structured, proximate, informed, voluntary, revocable, and recorded consent process before a high-stakes AI decision has consequences.
Ethics firewall. An enforceable governance barrier that blocks prohibited uses, not merely advises caution.
Graceful degradation. The ability of a system to move into safer, slower, less automated modes instead of collapsing or requiring full shutdown.
Human in the loop. A common phrase that only becomes meaningful when the human has time, understanding, authority, and protection to intervene.
Intervention ladder. A graduated sequence of controls from pause to restriction, human confirmation, isolation, safe mode, and full shutdown.
Machine-speed oversight. Monitoring and escalation capable of detecting machine-speed patterns quickly enough for human judgment to act.
Synthetic authority. The institutional weight given to machine-produced text, scores, recommendations, summaries, or rankings.
Trust infrastructure. The policies, records, rights, and habits that make AI adoption legitimate and governable.

Appendix L: Endnotes and Sources

These endnotes identify the public frameworks, laws, investigations, and scholarship supporting the arguments of each chapter, and they separate documented material from the labeled composite cases and speculative scenes. They are keyed by chapter and theme so the text stays readable while every substantive claim remains checkable. The research spine: the NIST AI Risk Management Framework and Generative AI Profile, the EU Artificial Intelligence Act, the OECD AI Principles, the Council of Europe Framework Convention on AI, the UNESCO Recommendation on the Ethics of AI, WHO guidance on AI for health, the NASA ASRS reporting model, SEC circuit-breaker practice, NIST SP 800-61 incident response, and AI incident-database research.

Author's Note

1. On the public record of AI incidents: AI Incident Database, Responsible AI Collaborative, incidentdatabase.ai—a public collection of documented AI harms and near misses.

2. On safety engineering lineage (checklists, incident reporting): A. Gawande, The Checklist Manifesto (Metropolitan Books, 2009); NASA Aviation Safety Reporting System (ASRS), asrs.arc.nasa.gov.

3. On composite and illustrative cases in governance writing: the practice follows disclosure norms recommended for nonfiction case material; all composites in this book are labeled in the text.

Preface

4. On AI systems as testable, auditable, and manageable: National Institute of Standards and Technology, Artificial Intelligence Risk Management Framework (AI RMF 1.0), NIST AI 100-1 (January 2023)—Govern, Map, Measure, Manage functions.

5. On documented model failure modes (hallucination, misclassification): NIST AI RMF 1.0, §Appendix B (AI risks); Stanford HAI, AI Index Report (annual), responsible-AI chapters.

6. On automation and human reliance: R. Parasuraman and V. Riley, "Humans and Automation: Use, Misuse, Disuse, Abuse," Human Factors 39, no. 2 (1997): 230–253.

7. On engagement-optimized design and compliance incentives: U.S. Federal Trade Commission, Bringing Dark Patterns to Light (staff report, September 2022).

8. On limits equal to inventions (regulatory history): D. Vogel, The Politics of Precaution (Princeton University Press, 2012), on how safety regimes follow technological harms.

Introduction

9. On alignment-centric framing of AI risk: S. Russell, Human Compatible: Artificial Intelligence and the Problem of Control (Viking, 2019).

10. On automation bias in clinical and administrative settings: K. Goddard, A. Roudsari, and J. C. Wyatt, "Automation Bias: A Systematic Review," Journal of the American Medical Informatics Association 19, no. 1 (2012): 121–127.

11. On the ironies of supervisory control: L. Bainbridge, "Ironies of Automation," Automatica 19, no. 6 (1983): 775–779.

12. On regulation arriving after harm (fire, chemicals, atoms): C. Perrow, Normal Accidents: Living with High-Risk Technologies (Basic Books, 1984; Princeton ed. 1999).

13. On oversight tempo versus system tempo: U.S. Securities and Exchange Commission and CFTC, Findings Regarding the Market Events of May 6, 2010 (joint report, September 30, 2010)—the "Flash Crash" analysis.

14. On human oversight as a legal requirement: Regulation (EU) 2024/1689 (EU Artificial Intelligence Act), Article 14 (human oversight) and Article 26 (deployer obligations).

15. On risk-based obligations for high-risk AI: EU AI Act, Articles 9 (risk management) and 12 (record-keeping/logging).

16. On intervention as a governance capacity: NIST AI RMF 1.0, Manage function (MG subcategories on incident response, decommissioning, and override).

Prologue

17. The Prologue is a speculative scenario, not a prediction or a description of any existing system. Brain-computer interface research and neural-data products referenced in later chapters are cited there.

Chapter 1: The Machine Is Not the Hardest Part

18. On evaluation, red-teaming, and capability gating: NIST AI RMF 1.0, Measure and Manage functions; NIST, Generative Artificial Intelligence Profile, NIST AI 600-1 (July 2024).

19. On audit trails and logging duties: EU AI Act, Article 12 (automatically generated logs for high-risk systems).

20. On rollback, throttling, and staged deployment as ordinary safety engineering: ISO/IEC 42001:2023, Artificial intelligence management system—requirements; ISO/IEC 23894:2023, AI risk management guidance.

21. On circuit breakers as reversible halts: U.S. SEC, "Investor Bulletin: Measures to Address Market Volatility"—market-wide circuit breakers and Limit Up–Limit Down mechanisms adopted after May 6, 2010.

22. On the Knight Capital deployment failure (uncontrolled automated trading, August 1, 2012): U.S. SEC, In re Knight Capital Americas LLC, Admin. Proc. File No. 3-15570 (October 16, 2013).

23. On automation complacency in human-machine teams: R. Parasuraman and D. H. Manzey, "Complacency and Bias in Human Use of Automation," Human Factors 52, no. 3 (2010): 381–410.

24. On organizational pressure overriding warnings: Columbia Accident Investigation Board, Report Volume I (NASA, August 2003), chap. 7–8 (organizational causes).

25. On the human-plus-system unit of analysis: D. D. Woods and E. Hollnagel, Joint Cognitive Systems: Foundations of Cognitive Systems Engineering (CRC Press, 2005).

Chapter 2: The Uncontrollable Human

26. On drift from careful review to routine acceptance: L. J. Skitka, K. L. Mosier, and M. Burdick, "Does Automation Bias Decision-Making?" International Journal of Human-Computer Studies 51, no. 5 (1999): 991–1006.

27. On default effects and choice architecture: R. H. Thaler and C. R. Sunstein, Nudge: Improving Decisions About Health, Wealth, and Happiness (Yale University Press, 2008).

28. On interface designs that make refusal costly: FTC, Bringing Dark Patterns to Light (2022); European Data Protection Board, Guidelines 03/2022 on deceptive design patterns in social media platform interfaces.

29. On engagement metrics as monetized attention: T. Wu, The Attention Merchants (Knopf, 2016).

30. On systems shaping the people inside them: S. Dekker, The Field Guide to Understanding 'Human Error', 3rd ed. (CRC Press, 2014).

31. On incentive design and safety culture: J. Reason, Managing the Risks of Organizational Accidents (Ashgate, 1997).

Chapter 3: The Human Becomes the System

32. On the Challenger launch decision and normalized deviance: D. Vaughan, The Challenger Launch Decision (University of Chicago Press, 1996); Report of the Presidential Commission on the Space Shuttle Challenger Accident (Rogers Commission, 1986).

33. On the Boeing 737 MAX design and oversight failures: U.S. House Committee on Transportation and Infrastructure, Final Committee Report: The Design, Development and Certification of the Boeing 737 MAX (September 2020).

34. On accountability when decisions are co-produced: EU AI Act, Article 26 (deployer obligations) and Article 86 (right to explanation of individual decision-making for certain high-risk systems).

35. On decision traceability requirements: EU AI Act, Articles 11–12 (technical documentation and logs); ISO/IEC 42001:2023, controls on AI system documentation.

36. On automated state decision systems harming citizens at scale: Royal Commission into the Robodebt Scheme, Report (Commonwealth of Australia, July 2023).

37. On the Dutch childcare benefits scandal (algorithmic risk scoring): Parliamentary Interrogation Committee on Childcare Allowance, Unprecedented Injustice (Ongekend onrecht), Dutch House of Representatives (December 2020).

38. On risk-scoring opacity in justice contexts: J. Angwin et al., "Machine Bias," ProPublica (May 23, 2016), and subsequent scholarly debate on COMPAS.

39. On institutional cognition and workflow-embedded influence: Canada, Directive on Automated Decision-Making (Treasury Board, 2019, as amended)—algorithmic impact assessment tiers and human-in-the-loop requirements.

Chapter 4: Identity Drift

40. On synthetic media and voice cloning capabilities: NIST AI 600-1 (Generative AI Profile), risks of synthetic content; U.S. FTC, "FTC Proposes New Protections to Combat AI Impersonation of Individuals" (February 2024).

41. On deepfake disclosure duties: EU AI Act, Article 50 (transparency obligations for certain AI systems, including synthetic content labeling).

42. On AI-generated text in professional communication: J. Hancock, M. Naaman, and K. Levy, "AI-Mediated Communication," Journal of Computer-Mediated Communication 25, no. 1 (2020): 89–100.

43. On style homogenization by writing assistants: M. Jakesch et al., "Co-Writing with Opinionated Language Models Affects

What Users Write," CHI Conference on Human Factors in Computing Systems (2023).

44. On authorship and provenance standards: Coalition for Content Provenance and Authenticity (C2PA), Content Credentials technical specification, c2pa.org.

45. On watermarking research and limits: NIST AI 600-1, provenance and watermarking discussion; partnership guidance from Partnership on AI, Responsible Practices for Synthetic Media (2023).

46. On the right to know when one is interacting with AI: EU AI Act, Article 50(1); OECD AI Principles (2019, updated 2024), transparency and explainability principle.

Chapter 5: Attention Is the New Control Surface

47. On the economics of captured attention: T. Wu, The Attention Merchants (Knopf, 2016); S. Zuboff, The Age of Surveillance Capitalism (PublicAffairs, 2019).

48. On recommender systems and engagement optimization: OECD, Recommender Systems and Their Effects (OECD Digital Economy Papers); EU Digital Services Act, Regulation (EU) 2022/2065, Articles 27 and 38 (recommender transparency and non-profiling option for VLOPs).

49. On persuasive technology design: B. J. Fogg, Persuasive Technology: Using Computers to Change What We Think and Do (Morgan Kaufmann, 2003).

50. On metric fixation in institutions: J. Z. Muller, The Tyranny of Metrics (Princeton University Press, 2018).

51. On alert fatigue and attention budgets in clinical settings: Agency for Healthcare Research and Quality (AHRQ), Patient Safety Network primer, "Alert Fatigue."

52. On market-wide automated halts protecting deliberation time: SEC-CFTC Flash Crash report (2010); NYSE and Nasdaq Limit Up–Limit Down plan documentation.

53. On information integrity as a governance objective: UNESCO, Guidelines for the Governance of Digital Platforms (2023).

Chapter 6: The Kill Switch Problem

54. On graduated, reversible market halts: SEC, market-wide circuit breaker rules (post-2010 reforms); Limit Up–Limit Down National Market System Plan.

55. On incident response discipline: NIST SP 800-61 Rev. 2, Computer Security Incident Handling Guide; forthcoming Rev. 3 public drafts.

56. On safe states and graceful degradation in safety-critical systems. IEC 61508, Functional safety of electrical/electronic/programmable electronic safety-related systems (safe state concepts).

57. On separation of duties for critical actions: ISO/IEC 27001:2022, Annex A control on segregation of duties; nuclear "two-person rule" practice, U.S. DoD and NRC documentation.

58. On multi-party authorization in high-consequence systems: International Atomic Energy Agency (IAEA), safeguards and physical protection guidance (INFCIRC/225/Rev.5).

59. On evidence preservation during incidents: NIST SP 800-86, Guide to Integrating Forensic Techniques into Incident Response.

60. On rehearsed shutdown and restart procedures: FAA and ICAO operational drill requirements; NTSB accident-investigation methodology as a model for post-incident learning.

61. On the EU AI Act's stop conditions: EU AI Act, Article 14(4)(e) —human overseers must be able to intervene or interrupt the system through a "stop" button or similar procedure.

Chapter 7: Consent as a Ritual

62. On informed consent doctrine: Faden and Beauchamp, A History and Theory of Informed Consent (Oxford University Press, 1986).

63. On the collapse of click-through consent: FTC, Bringing Dark Patterns to Light (2022); D. Solove, "Murky Consent: An Approach to the Fictions of Consent in Privacy Law," Boston University Law Review 104 (2024).

64. On GDPR consent standards (freely given, specific, informed, unambiguous): Regulation (EU) 2016/679 (GDPR), Articles 4(11) and 7; EDPB Guidelines 05/2020 on consent.

65. On rights regarding automated decisions: GDPR, Article 22 (decisions based solely on automated processing); EU AI Act, Article 86.

66. On dark patterns in consent flows: EDPB Guidelines 03/2022; California Privacy Rights Act regulations on dark patterns.

67. On workplace algorithmic management and consent limits: EU Platform Work Directive (2024) provisions on algorithmic management; ILO reports on algorithmic management.

68. On public-sector transparency for automated decisions: UK Algorithmic Transparency Recording Standard (Central Digital and Data Office); Canada, Directive on Automated Decision-Making.

69. On hiring-tool notice and audit duties: New York City Local Law 144 of 2021 (automated employment decision tools—bias audit and candidate notice).

70. On consent for minors and families: UN Convention on the Rights of the Child, General Comment No. 25 (2021) on children's rights in the digital environment.

Chapter 8: Oversight at Machine Speed

71. On post-market monitoring and serious-incident reporting: EU AI Act, Articles 72–73 (post-market monitoring; reporting of serious incidents).

72. On AI incident collection and learning: AI Incident Database; OECD AI Incidents Monitor (AIM).

73. On aviation's non-punitive reporting model: NASA ASRS program documentation; ICAO Annex 13 (accident and incident investigation).

74. On healthcare incident learning systems: Institute of Medicine, To Err Is Human (National Academies Press, 2000); WHO patient-safety incident reporting guidance.

75. On continuous model monitoring and drift: NIST AI RMF 1.0, Measure function; ISO/IEC 23894:2023.

76. On audit and assurance ecosystems for AI: UK Department for Science, Innovation and Technology, AI assurance roadmap materials; IAAA and algorithmic-audit literature.

77. On time-bounded disclosure norms: U.S. SEC cybersecurity disclosure rule (2023)—material incident disclosure on Form 8-K within four business days—as a tempo benchmark.

78. On dissent channels in technical organizations: CAIB Report (2003) recommendations on minority opinion and independent technical authority at NASA.

Chapter 9: Cognitive Rights

79. On neurorights in national law: Chile, Law No. 21.383 (2021), constitutional amendment protecting cerebral activity and mental integrity; Chilean Supreme Court, Girardi v. Emotiv (2023).

80. On neural data as sensitive personal data: Colorado H.B. 24-1058 (2024), amending the Colorado Privacy Act to cover neural data; California S.B. 1223 (2024), amending the CCPA.

81. On international neurotechnology ethics: UNESCO, Recommendation on the Ethics of Neurotechnology (adopted November 2025); OECD, Recommendation on Responsible Innovation in Neurotechnology (2019).

82. On cognitive liberty scholarship: N. A. Farahany, The Battle for Your Brain: Defending the Right to Think Freely in the Age of Neurotechnology (St. Martin's Press, 2023).

83. On mental integrity and human rights framing: Council of Europe, Framework Convention on Artificial Intelligence and Human Rights, Democracy and the Rule of Law (CETS No. 225, opened for signature September 5, 2024).

84. On subliminal and manipulative techniques: EU AI Act, Article 5(1)(a)–(b)—prohibited practices including manipulative or exploitative techniques causing significant harm.

85. On children's cognitive protection: EU AI Act, Article 5 protections; UNCRC General Comment No. 25 (2021).

86. On the right to explanation and appeal: GDPR Article 22; EU AI Act Article 86; Council of Europe Framework Convention, remedies provisions.

Chapter 10: Governance Beyond Borders

87. On the first binding international AI treaty: Council of Europe Framework Convention on AI (CETS No. 225, 2024)—signatories include the EU, United States, United Kingdom, and others.

88. On intergovernmental principles: OECD AI Principles (2019; updated May 2024); G7 Hiroshima Process International Code of Conduct for Organizations Developing Advanced AI Systems (2023).

89. On UN-level governance: UN General Assembly Resolution A/78/L.49, "Seizing the Opportunities of Safe, Secure and Trustworthy AI" (March 2024); UN High-Level Advisory Body on AI, Governing AI for Humanity (final report, September 2024).

90. On summit diplomacy: Bletchley Declaration (UK AI Safety Summit, November 2023); Seoul Declaration (May 2024); Paris AI Action Summit outcomes (February 2025).

91. On registries and databases of high-risk systems: EU AI Act, Article 71 (EU database for high-risk AI systems).

92. On aviation as the model for cross-border safety reciprocity: Convention on International Civil Aviation (Chicago Convention, 1944) and ICAO standards-and-recommended-practices system.

93. On nuclear verification as precedent for inspections: Treaty on the Non-Proliferation of Nuclear Weapons (1968); IAEA comprehensive safeguards agreements and Additional Protocol.

94. On cross-border data and AI supply chains: OECD, "Mapping Data Flows and AI Value Chains" working papers; WTO Joint Statement Initiative on e-commerce discussions.

95. On procurement as governance leverage: U.S. OMB Memorandum M-24-10 (March 2024) on agency AI governance and M-24-18 on AI acquisition.

Chapter 11: The Intelligence Arms Race

96. On military AI decision-support and autonomy debates: UN Group of Governmental Experts on Lethal Autonomous Weapons Systems (CCW GGE LAWS), reports 2019–2024; ICRC position on autonomous weapon systems (2021).

97. On political declarations for responsible military AI: U.S. State Department, Political Declaration on Responsible Military Use of Artificial Intelligence and Autonomy (2023, 50+ endorsing states).

98. On national strategy competition: U.S. Executive Order 14110 on Safe, Secure, and Trustworthy AI (October 2023) and successor policies; national AI strategies tracked by OECD.AI policy observatory.

99. On speed advantages in markets: M. Lewis, Flash Boys (W. W. Norton, 2014); SEC-CFTC Flash Crash report (2010).

100. On compute and capability concentration: Stanford HAI AI Index (annual), industry-versus-academia frontier model analyses.

101. On frontier-model safety evaluation regimes: UK AI Safety (now Security) Institute and U.S. AI Safety Institute publications and evaluation frameworks (2024–2025); International AI Safety Report (2025), chaired by Y. Bengio.

102. On tempo limits as racing rules: market circuit breakers (SEC); human-decision requirements in EU AI Act Article 14; nuclear command-and-control "human in the loop" policy literature.

103. On verification approaches for AI agreements: M. Baker et al., "Verification Methods for International AI Agreements" (arXiv, 2024) and related governance scholarship.

Chapter 12: The Ethics Firewall

104. On prohibited AI practices as red lines: EU AI Act, Article 5 (unacceptable-risk practices, including social scoring by public authorities and certain biometric uses).

105. On institutional ethics infrastructure: ISO/IEC 42001:2023 (AI management systems); IEEE 7000-2021, Model Process for Addressing Ethical Concerns During System Design.

106. On professional codes with teeth: ACM Code of Ethics and Professional Conduct (2018); medical-profession licensure as an enforcement model.

107. On layered defenses in safety engineering: J. Reason's "Swiss cheese" model, in Human Error (Cambridge University Press, 1990).

108. On organizational capture and waiver-driven erosion: D. Vaughan, The Challenger Launch Decision (1996)—normalization of deviance.

109. On technical safety nets that hold under pressure: EU AI Act Article 14(4)(e) stop conditions; NIST AI 600-1 on model-level safeguards.

110. On whistleblower protection as a firewall component: EU Whistleblower Directive (EU) 2019/1937; U.S. SEC whistleblower program as institutional design references.

111. On ethics review that binds decisions: WHO, Ethics and Governance of Artificial Intelligence for Health (2021, updated LMM guidance 2024–2025).

Chapter 13: The Post-Intervention Future

112. On public-sector AI registers: Amsterdam and Helsinki municipal AI registers (launched 2020); EU AI Act Article 71 database.

113. On operationalizing trustworthy AI in agencies: U.S. OMB M-24-10 (minimum risk-management practices for rights- and safety-impacting AI); UK Algorithmic Transparency Recording Standard.

114. On governance operating models in firms: ISO/IEC 42001:2023; NIST AI RMF Playbook (companion resource).

115. On human oversight as ongoing duty, not launch gate: EU AI Act Articles 14 and 72 (oversight plus post-market monitoring).

116. On AI literacy obligations: EU AI Act, Article 4 (AI literacy of providers' and deployers' staff).

117. On teaching system behavior before deployment to children: UNESCO, AI competency frameworks for students and teachers (2024).

118. On incident learning as civic memory: OECD AI Incidents Monitor; AI Incident Database.

Chapter 14: The Power of Intervention

119. The closing chapter synthesizes frameworks cited earlier; primary anchors: NIST AI RMF 1.0 (Govern-Map-Measure-Manage); EU AI Act Articles 5, 9, 12, 14, 26, 50, 72–73, 86; OECD AI Principles (2024 update); Council of Europe Framework Convention (CETS No. 225).

120. On responsibility remaining human: Council of Europe Framework Convention, accountability and remedies provisions; OECD Principle on accountability.

121. On commitments-with-owners as governance practice: ISO/IEC 42001 role-assignment requirements; NIST AI RMF Govern function (roles, responsibilities, accountability structures).

Epilogue

122. On drills over documents: aviation recurrent-training requirements (FAA Part 121; EASA equivalents) as the rehearsal model.

123. On leaving successors better options: UN, Governing AI for Humanity (2024), intergenerational framing of AI governance.

124. On the oath form in professional life: the Hippocratic tradition and modern engineering-ethics oaths (e.g., Canadian "Iron Ring" Obligation) as civic-commitment precedents.

References and Further Reading

• National Institute of Standards and Technology. Artificial Intelligence Risk Management Framework (AI RMF 1.0).

- National Institute of Standards and Technology. Generative AI Profile for the AI Risk Management Framework.

- European Union. Regulation (EU) 2024/1689, Artificial Intelligence Act.

- Organisation for Economic Co-operation and Development. OECD AI Principles.

- Council of Europe. Framework Convention on Artificial Intelligence and Human Rights, Democracy and the Rule of Law.

- UNESCO. Recommendation on the Ethics of Artificial Intelligence.

- World Health Organization. Ethics and Governance of Artificial Intelligence for Health.

- ISO/IEC 42001. Artificial intelligence management system standard.

- ISO/IEC 23894. Artificial intelligence risk management guidance.

- IEEE 7000 series. Standards and guidance on ethically aligned design and system governance.

- European Union. General Data Protection Regulation (GDPR), Regulation (EU) 2016/679.

- AI Incident Database. Research and public incident collection on AI harms and near misses.

- Partnership on AI. Responsible practices, incident reporting, and human-centered AI governance resources.

- ACM FAccT Conference Proceedings. Research on fairness, accountability, transparency, and sociotechnical systems.

- Cathy O'Neil. Weapons of Math Destruction.

- Frank Pasquale. The Black Box Society.

- Lawrence Lessig. Code and Other Laws of Cyberspace.

- Charles Perrow. Normal Accidents.

- Diane Vaughan. The Challenger Launch Decision.

- Sidney Dekker. The Field Guide to Understanding Human Error.

These sources form the research spine for the book's governance architecture.

Acknowledgments

A book about intervention is, fittingly, the product of many people who were willing to interrupt me. I am grateful to the colleagues, readers, advisors, and institutional leaders whose questions shaped the argument. I am also grateful to the builders, teachers, clinicians, engineers, auditors, regulators, and parents who work every day to keep human judgment alive inside complex systems.

Most of all, this book is dedicated to the next generation, who will inherit not only the tools we build, but the limits we had the courage to design around them.

About the Author

Dr. Najib M. Najib, PhD, is an executive, author, and governance-focused thought leader working at the intersection of technology, institutional risk, leadership, and human agency. His work explores how organizations and societies can preserve human judgment as artificial intelligence becomes embedded in decision-making systems, public institutions, and everyday life.

In Power of Intervention, Dr. Najib develops a framework for cognitive sovereignty and human-centered AI governance, urging leaders to build systems that preserve consent, identity, accountability, and the right to intervene.

About the Superhumans Rise Series

Power of Intervention is Book One of the Superhumans Rise series from Superhumans Rise Publishing—books for leaders, builders, educators, and citizens determined to become more capable without becoming less human.

Each volume in the series pairs clear-eyed foresight about a rising technology with practical frameworks readers can put to work immediately—in boardrooms, classrooms, hospitals, agencies, and homes. The premise is the same throughout: human potential and human governance must rise together. Capability without stewardship is not elevation. It is exposure.

Future volumes will extend the series into the disciplines where human and machine capability are converging fastest—leadership, learning, health, and civic life—each built on the standard this book sets: name the risk honestly, give the reader tools, and keep the human answerable.

Ideas that Elevate. Principles that Endure.

www.ingramcontent.com/pod-product-compliance
Lightning Source LLC
LaVergne TN
LVHW090258210926
848480LV00001B/65
9798951754004